The Etl

Edited by

Seana Moran
Clark University, USA

David Cropley
University of South Australia, Australia

James C. Kaufman
University of Connecticut, USA

First published 2014 by
PALGRAVE MACMILLAN

Palgrave Macmillan in the UK is an imprint of Macmillan Publishers Limited, registered in England, company number 785998, of Houndmills, Basingstoke, Hampshire RG21 6XS.

Palgrave Macmillan in the US is a division of St Martin's Press LLC, 175 Fifth Avenue, New York, NY 10010.

Palgrave Macmillan is the global academic imprint of the above companies and has companies and representatives throughout the world.

Palgrave® and Macmillan® are registered trademarks in the United States, the United Kingdom, Europe and other countries.

ISBN 978–1–137–33353–7 hardback
ISBN 978–1–137–33352–0 paperback

This book is printed on paper suitable for recycling and made from fully managed and sustained forest sources. Logging, pulping and manufacturing processes are expected to conform to the environmental regulations of the country of origin.

A catalogue record for this book is available from the British Library.

A catalog record for this book is available from the Library of Congress.

The Ethics of Creativity

For BK

SM

For MJC

DHC

*For Ron Beghetto: friend and colleague,
and one of the most creative and ethical people I know!*

JCK

Contents

Figures and Tables

Figures

Tables

Acknowledgments

The editors would like to thank Nicola Jones, Harriet Barker, Maryam Rutter, Sally Osborn, Linda Auld, and Libby Forrest at Palgrave Macmillan. Seana Moran would like to thank Vera John-Steiner and Howard Gardner, who encouraged her interest in the system dynamics of creativity, collaboration, and morality; and she thanks all the authors of this volume who are helping encourage that same interest in others. Her studies cited in Chapter 16 were funded in part by the American Association of University Women and the Edmond J. Safra Center for Ethics at Harvard University. James Kaufman would like to thank his new colleagues at the University of Connecticut and (as always) Allison, Jacob, and Asher. Michael Mumford and colleagues would like to thank Cheryl Stenmark, Alison Antes, Steven Murphy, and Chase Thiel for their contributions to the work described in Chapter 15, and they acknowledge that parts of that work were supported by grants from the National Science Foundation and the National Institutes of Health, Michael D. Mumford, Principal Investigator.

This material in chapter 9 is based upon work supported by the U.S. Department of Homeland Security under Grant Award Number 2010-ST-061-RE0001. The views and conclusions contained in this chapter are those of the authors and should not be interpreted as necessarily representing the official policies, either expressed or implied, of the U.S. Department of Homeland Security.

Notes on Contributors

Mark Coeckelbergh teaches philosophy in the Philosophy Department at the University of Twente, The Netherlands, and is managing director of the 3TUCentre for Ethics and Technology. He is the author of *Liberation and Passion, The Metaphysics of Autonomy, Imagination and Principles, Growing Moral Relations, Human Being @ Risk,* and numerous articles in the ethics of technology in information technologies and robotics, medicine and healthcare, and related to the environment.

Arthur Cropley received his PhD from the University of Alberta in 1965. He taught at universities in Australia, Canada, and Germany, retiring in 1998. He then worked as an adjunct professor at the University of Latvia for 12 years, from which he received an honorary doctorate in 2005. He has published 27 books and has received several awards from international associations. In 2008, he was made an Officer of the Order of the Three Stars by the President of Latvia.

David H. Cropley is Deputy Director of the Defence and Systems Institute (DASI) and Associate Professor of Engineering Innovation. He completed a PhD in measurement systems engineering at the same institution in 1997, and a Graduate Certificate in Higher Education from the Queensland University of Technology in 2002. His research interests lie in systems engineering, creativity and innovation in engineering processes, and the nexus of creative problem-solving and engineering. In 2013, he co-authored *Creativity and Crime: A Psychological Analysis.*

Mihaly Csikszentmihalyi is the C.S. and D.J. Davidson Professor of Psychology at the School of Behavioral and Organizational Sciences and the Peter F. Drucker Graduate School of Management, and co-Director of the Quality of Life Research Center. He is also emeritus professor of human development at the University of Chicago and holds honorary doctor of science degrees from several universities. He is a fellow of the American Academy of Arts and Sciences, the American Academy of Political and Social Science, and a foreign member of the Hungarian Academy of Sciences.

Charlotte Dixon is a doctoral candidate at the Harvard Graduate School of Education. Her research interests in the sociocultural psychology of artistic creativity grew from her work as a documentary and press photographer. She has an MFA from the Yale University School of Art. She teaches both still- and time-based media at Maine Media College and works with regional museums and schools to develop visual arts and culture curricula for K-12 educators.

Anthony Finn has led 15 collaborative international research programs with universities, government laboratories, and research institutes from Australia, the USA, the UK, Singapore, Malaysia, and Indonesia. He graduated from Cambridge University in 1988. He has published two books and ninety book chapters, journal articles, conference papers, and research reports. He is on the editorial board of four academic journals and has received several international awards for scientific achievement and innovation. His research interests largely focus on autonomous and unmanned vehicle systems and their applications.

Howard Gardner trained in developmental psychology and neuropsychology; he has studied and written extensively about intelligence, creativity, leadership, the arts, and professional ethics. Recent books include *Good Work, Changing Minds, The Development and Education of the Mind, Multiple Intelligences: New Horizons*, and *Truth, Beauty, and Goodness Reframed*. His latest co-authored book, *The App Generation: How Today's Youth Navigate Identity, Intimacy, and Imagination in a Digital World*, was published in October 2013.

Jack A. Goncalo received his PhD in Business Administration from the University of California, Berkeley. His research on creativity highlights the importance of individualism for sparking creative thought and for facilitating the free exchange of ideas in groups. His work spanning the fields of management and psychology has been published in *Organizational Behavior and Human Decision Processes, Management Science, Journal of Experimental Social Psychology, Journal of Experimental Psychology: General*, and *Psychological Science*. He co-edited the book *Research on Managing Groups and Teams: Creativity in Groups*.

Daniel J. Harris is a doctoral candidate at the University of Nebraska Omaha. He is in the Industrial Organizational Psychology program. His research focuses on malevolent creativity, destructive leadership, and their personality-based antecedents.

Helen Haste is Professor Emerita of Psychology at the University of Bath. She is also a visiting professor at the University of Exeter and at the Harvard Graduate School of Education. Her research interests lie in the development of moral, social, and political values and citizenship; the role of metaphor in cognition, dialogue, and culture; and the relationship between science and society, particularly media and cultural images of science. She is a fellow of the Academy for Social Sciences in the UK.

Reuben Hersh is Professor Emeritus of Mathematics at the University of New Mexico. He is the co-author (with Philip J. Davis) of *The Mathematical Experience* and *Descartes Dream*, and of *What Is Mathematics, Really?* and *Loving and Hating Mathematics* (with Vera John-Steiner). His writings include many articles and some mathematical poetry.

Vera John-Steiner is Regents' Professor Emerita of Education and Linguistics at the University of New Mexico. Her research includes creativity, collaboration, cultural historical theory, and psycholinguistics. She co-edited Vygotsky's *Mind in Society*, an influential text in the human sciences. In *Notebooks of the Mind*, which received the William James Award in 1990, she explored the diversity of thought and creative endeavors. In *Creative Collaboration*, she documented the impact of working partnerships. Recently, she co-authored *Loving and Hating Mathematics* with Reuben Hersh. She has received many honors and has taught and lectured in Latin America, Europe, and the United States.

James C. Kaufman is Professor of Educational Psychology at the University of Connecticut, researching creativity. Previously, he taught at the California State University, San Bernardino, where he directed the Learning Research Institute. He received his PhD from Yale University in 2001. He has written and edited 24 books, including *Creativity 101*, *The Cambridge Handbook of Creativity*, *Essentials of Creativity Assessment*, *The International Handbook of Creativity*, and *The Psychology of Creative Writing*. He is the Series Editor of the Psych 101 series from Springer.

Qin Li is a doctoral candidate at Claremont Graduate University. Her field of research is positive developmental psychology, aiming for enhanced understanding of the creative process in order to develop methods of creativity enhancement. She examines creativity in relation to talent development, affect, mental illness, and expertise. She has presented her work at the American Psychological Association Conference, the Western Psychological Conference, and the Society for Research in Child Development.

Gina Scott Ligon is Assistant Professor at the Department of Psychology, University of Villanova. Her research focuses on creativity and leadership, with a particular emphasis on the intersection of malevolent creativity and destructive leadership in extreme ideological organizations.

Alexandra E. MacDougall is a doctoral candidate at the University of Oklahoma. She is in the industrial and organizational psychology program and her research interests include ethics and leadership.

Seana Moran is Research Assistant Professor at the Department of Psychology, Clark University. She received her doctorate in human development and psychology from Harvard University. Her research focuses on the intersections of creativity, morality/ethics, life purpose, and wisdom as individuals strive to contribute to their communities. She co-edited *Multiple Intelligences Around the World* and several volumes of the *Creative Classrooms* series, and co-authored *Creativity and Development*. She has published numerous articles and received several grants, awards, and fellowships.

Kellen Mrkva is a doctoral candidate at the University of Colorado—Boulder. He is in the graduate program for social psychology and studies morality, emotion, and decision-making with his advisor Leaf Van Boven. He completed his undergraduate work at the University of Notre Dame, where he worked with Darcia Narvaez, studying moral identity, moral judgment, and charitable donations.

Michael Mumford is the George Lynn Cross Distinguished Research Professor of Psychology and he directs the Center for Applied Social Research. He received his doctoral degree from the University of Georgia in 1983 in industrial and organizational psychology and psychometrics. He is a fellow of the American Psychological Association (Divisions 3, 5, 10, 14), the Society for Industrial and Organizational Psychology, and the American Psychological Society. He has written more than 300 articles on creativity, leadership, ethics, and planning.

Darcia Narvaez studies moral development with a particular focus on early-life effects of the evolved developmental niche on the neurobiology underpinning moral functioning. She has developed interventions for moral character development, including integrating moral character skill development into academic instruction. She emphasizes "moral complexity" and the importance of both deliberative and intuitive processes in ethical expertise. She has co-authored or co-edited eight books. Her latest authored book is *The Neurobiology and Development of Human Morality: Evolution, Culture and Wisdom.* She is editor of the *Journal of Moral Education.*

James Noonan is a doctoral candidate at the Harvard Graduate School of Education. His research addresses effective learning environments for teachers. He worked for eight years at a Boston-based non-profit organization where he developed curricula and teacher training programs aimed at developing the civic and social competencies of young people.

David R. Peterson is a doctoral candidate at the University of Oklahoma. He is in the industrial and organizational psychology program. His research interests include creativity, leadership, and ethics.

Roni Reiter-Palmon is Issacson Professor of I/O Psychology and Director of the I/O Psychology graduate program at the University of Nebraska at Omaha (UNO). Her research focuses on the cognitive stages of creativity as well as the influence of personality and individual differences on creativity and leadership.

Ruth Richards is Professor at the School of Psychology and Interdisciplinary Inquiry, Saybrook University. She has studied creativity in educational, clinical, social action, and spiritual contexts, authored numerous papers, and edited or contributed to two books: *Eminent Creativity, Everyday Creativity, and Health* (with Mark Runco) and *Everyday Creativity and New Views of Human Nature.* In 2009, she won the Arnheim Award for Lifetime Achievement from Division 10 of the American Psychological Association. She examines whether creative process can bring people to greater health, offering new ways to be present with themselves, each other, and life's possibilities.

Robert J. Sternberg received his PhD from Stanford University and holds 13 honorary doctorates. He is the President of the University of Wyoming as well as President of the Federation of Associations in Behavioral and Brain Sciences and a past President of the American Psychological Association. He is the author of close to 1,500 articles, books, and book chapters. He was a professor for 30 years at Yale University, and also has served as Provost at Oklahoma State University and the Dean of Arts and Sciences at Tufts University.

Kirsi Tirri is Professor of Education and Research Director of the Department of Teacher Education, University of Helsinki, Finland, as well as a visiting scholar with the Stanford Center on Adolescence. She has been the President of the European Council for High Ability and the President of the International Studies Special Interest Group in the American Educational Research Association. Her research interests include moral and religious education, gifted education, teacher education, and cross-cultural studies.

Lynne C. Vincent is a postdoctoral researcher at the Owen Graduate School of Management, Vanderbilt University. She received her PhD in Organizational Behavior at Cornell University. Her research explores the relationships among creativity, morality, and ethics, especially addressing the effects of the creative identity on behavior and how creativity can mitigate the effects of negative experiences, such as social rejection. Her research has been published in the *Journal of Experimental Psychology: General* and *Psychological Science*.

Thomas A. Zeni is a doctoral candidate at the University of Oklahoma. Enrolled in the industrial and organizational psychology program, his research interests include leadership and ethics.

Introduction: The Crossroads of Creativity and Ethics

Seana Moran
Clark University, USA

"New is better." "Innovate faster." "Change the world."

"More creativity!" seems to be the current mantra for success. Institutions, cities, and nations seek globally for people who will "break the mold," "cultivate disruption," or "hack the future" to provide competitive advantage and "stay ahead of the curve." Several higher education institutions have augmented their traditional strengths of general and professional knowledge by promoting the "twenty-first-century skills" of creativity and collaboration.

Amid the frenzy, some skeptics have a clarion call for us: Where is this leading? How will it proceed? Who is watching the effects over time, such as who will benefit and who might be harmed? As news reports of subprime lending or overused antibiotics imply, what seems good in the short term can contribute to catastrophes later on. As discussions of pollution and climate change suggest, what is good for producers and direct users may have adverse consequences for communities at large. These questions and situations address ethics, which describes moral norms and codes of conduct that provide direction for behaving properly. Ethics applies to situations that involve relations among people or the effects that our behavior may have on others or some greater good.

Numerous recent examples of creative ideas or products let loose in society have resulted in destructive consequences, some which continue to spread. Financial derivatives contributed to the "great recession" starting in 2008, which destabilized jobs and financial markets worldwide. Social media services launched in the mid-2000s have spawned an accumulation of "Big Data," which is eroding privacy and increasing a culture of constant surveillance (Debatin et al., 2009). The jury is still out for hydraulic "fracking" and for genetic modification in medical therapies and food production. But the controversies revolve around whether we are being mindful or careless of these technologies' consequences regarding our water supply, food nutrition, and, ultimately, the health of ourselves, other living organisms, and future generations (Sandel, 2007). In short, how ethical is our creativity?

Other recent events signal that our ethical frameworks may require updating, if not transforming. Established rules based on old assumptions may not work any more. New computers and machines that kill can reduce the perceived human cost of warfare (or, later, perhaps law enforcement), because fewer soldiers or officers die in the line of duty. Yet, we are faced with ethical quandaries (see Finn, this volume): Will drones make military action more common and acceptable by making it "easy"? Who is responsible for the killings? How are they responsible? Similarly, internet-based classified listings sites or car-sharing sites or lodging rental sites have spawned a person-to-person commerce infrastructure that allows individuals to buy, sell, or rent various assets without institutional middlemen. These burgeoning opportunities bring up new ethical questions: How can "government" (whoever that may be) regulate when there are no clear mechanisms for enforcement? How are standards of quality, transparency, and monetary transactions to be maintained? If situations go awry, how will responsibility be allocated? In other words, should we intentionally think about ethics in creative ways to address the changing ways in which people can affect each other's wellbeing?

Creativity as novel, useful contributions to culture—is it "good"?

Most of the insightful, well-known scholars who contributed chapters to this volume start with the current standard definition of creativity as the introduction of a novel idea or product that is eventually deemed useful by a community to be widely used by the current generation and perhaps taught to future generations (Runco & Jaeger, 2012). Systems models of creativity (Csikszentmihalyi, 1988, 1999; Moran, 2009a, 2010e; Moran & John-Steiner, 2003; see also Noonan & Gardner and Moran, this volume) emphasize how creativity involves both a cognitive-emotional process of coming up with the novelty, *and* a social process that requires others' recognition and acceptance of the novelty, either through powerful, expert gatekeepers or social diffusion (Csikszentmihalyi, 1999; Rogers, 1983[1962]; Sosa, 2011; Sosa & Gero, 2004; Stein, 1993; Subotnik, Jarvin, Moga, & Sternberg, 2003).

There is evaluation and acceptance of the novelty on the part of other people: Copernicus, Einstein, Edison, Marie Curie, Martha Graham, Shakespeare, and Gandhi did not become the historically important individuals they are without others buying into their contributions. Creativity has tremendous power as a driving engine of cultural change (Glaveanu, 2011; Moran & John-Steiner, 2003; Valsiner, 2000): not only materially in a new product or in making money, but also culturally through changing assumptions and beliefs, and socially because once a critical mass of people accepts the novelty, the group is a force to be reckoned with. At creativity's most transformative impact, what was initially creative becomes the new norm. Although products that fit well with the existing social milieu are more easily adopted (Mumford & Gustafson, 1988), over time, radical creativity—and even the slower accumulation of smaller adaptations—can transform a culture's foundations (Moran & John-Steiner, 2003). For example, Einstein's idea of relativity is not just a physics concept any more; it also affects people's general worldviews that perspective and stance matter, and this notion has rippled into psychology, literature, movies, and law.

Despite the promise of creativity, we are ambivalent about, and even biased against, creativity (Moran, 2010c; Mueller, Melwani, & Goncalo, 2012; see also A. Cropley, this volume). We do not want non-stop creativity—incessant tax law changes, or airport screening technologies, or software update downloads, for instance, tend to irritate people. Furthermore, as consumers, most of us prefer reliability and safety, rather than creativity, in our commercial airline pilots or surgeons or farmers. Experimentation may be considered acceptable in the military (for pilots), clinical trials (for medicine), and agricultural research universities (for food; Moran, 2009a). We tend to prefer that creativity be put into "self-contained" endeavors removed from everyday life—like skunkworks, or labs, or test chambers—rather than directly into the mainstream culture (Jaques, 1955; Stacey, 1996).

This volume's authors address as their common purpose, but from different perspectives, the question: What is going on at the crossroads of creativity and ethics? History and everyday life show that creativity and ethics go hand in hand. Innovations to improve living conditions, a morally laudable aim, can stimulate far-reaching effects on social relations that alter the obligations of individuals to each other and the wider society. For example, the introduction of electricity and household appliances in the twentieth century improved the daily lives of working women who could afford the appliances, because they formerly had to work "double shifts" in the workplace and at home.

Yet, for upper-middle-class women, who previously sent their laundry to be cleaned, the impact was less positive because the new appliances made these tasks "do-it-yourself" chores (Tenner, 1996). The longer-term repercussions of electricity freeing up women's time, some argue, contributed to boredom and a search for new meaning (Friedan, 1963), women's liberation, entry into the paid workforce, and increased demand for childcare and other domestic support industries. The ethics of what women were supposed to be doing in society, and relationships of women to other cultural members, changed along with the technology.

Where do we start?

This volume organizes several perspectives that have broached the crossroads of creativity and ethics. When launching a new investigational arena, especially one that aims to integrate two formerly separate conceptual spaces, metaphors can help (see Moran, 2009b). Although metaphors are not theories, many creative thinkers start with metaphors to give structure to their thinking: Newton used the metaphor of the universe as a clock, Darwin of diversity as a tree, Einstein of a light beam as a train. The use of a familiar, concrete object or symbol with understood properties can aid the development of understanding the properties of the less familiar concept—and, importantly, create pathways to advance thought.

Given this volume's aim to stimulate further research at the crossroads of creativity and ethics, in this introduction I consider five metaphors for how to characterize this intersection: a magnet, a ripple, whirls of smoke, a map, and dough. The goal of this metaphor-based discussion is not to verify and document which metaphor is "right," but rather to provide scaffolding to stimulate thoughtfulness, perspective-taking, and wider horizons of possibilities for reading the chapters that follow. Which metaphor resonates the most with you? Disturbs you? Intrigues you? As you read different chapters, what metaphor(s) seem implicit in the authors' arguments? What are further implications that arise from using any of these metaphors as a basis for further research?

Magnet: Creativity *or* ethics

A magnet has two poles that attract and repel, creating fields and boundaries of influence around each pole. With this metaphor, ethics is viewed as one pole, representing stable rules that help people know what to approach and avoid in advance, such as "give the bigger half of the candy bar to the other person," "keep your promises," and "don't cheat." Creativity is viewed as the other pole, representing flux, change, and disruption. Creativity and ethics are separate domains. They do not directly interact. Individuals and groups are attracted to one or the other pole. If they prefer stability, they are pulled toward ethics; if they are more flexible or prefer change, they are pulled toward creativity. In scholarship in the mid-twentieth century, ethics thinkers and creativity thinkers similarly kept to their respective poles, scoping their work not to attract attention to the other.

Ethics governs social interactions, where people directly affect each other. Creativity reigns in the symbolic realm of ideas, artifacts, and meanings (Moran & John-Steiner, 2003). The symbolic realm only indirectly affects people's interactions, so it often is not considered within the purview of ethics. Creativity is often associated with play, self-expression, art, and theoretical sciences. These domains usually are not conceptualized as prosocial or moral—they are isolated, special, "lone genius" domains removed from the "everyday world" (cf. Hersh, 1990).

Since the magnet's two poles stay apart, creativity is viewed as amoral; the rules are at the other ethical pole, and thus do not apply at the creativity pole. At the ethics pole, much of the time, given that most people tend to be relatively loss avoidant (Kahneman, Slovic, & Tversky, 1989), ethics tends to reject the radically new, although adaptive novelties could be attractive if they are easy to use (Mumford & Gustafson, 1988). The psychological biases of behavioral economics reinforce the notion that most people prefer the stable ethics pole, as people tend to make decisions based on what they already recognize, what others around them do, or what is most available in their immediate environment, rather than through experimentation or radical departures from the status quo (Gigerenzer, 2008, 2010; Kahneman, Slovic, & Tversky, 1989). The creativity pole tends to require more effort because there is more uncertainty, and people must discern among the wider possibilities, which can be considered less efficient (Hirsh, Mar, & Peterson, 2012).

What does this metaphor of the magnet look like in real life? For example, Einstein, whose equations are the backbone of nuclear weapons, is not held morally responsible for the atomic bomb. He is placed on the creative pole. His symbolic creation that paved the way for unleashing energy in the atom is too far removed from the social impact of Hiroshima and Nagasaki. Similarly, Renaissance artists' invention of three-dimensional perspective in painting, and Picasso's flattening of three-dimensional perspective in the early twentieth century, are not considered moral acts, only creative acts. More recently, a news item told the story of a subway incident. One man suffered a seizure and fell onto the tracks. A train was coming. Another man jumped down to save the first man. Instead of pulling the man up, because he was convulsing, the hero laid on top of the other man and held him down as flat as possible until the train passed over them. This was described as a moral act, but not a creative one.

Although this metaphor still underpins much creativity research, this characterization of the creativity–morality interaction—which is no interaction at all—is unsatisfying. It sidesteps the issues. Why are Einstein and the painters not held responsible for the consequences of their work? Einstein himself recognized the repercussions of his equations (Butcher, 2005; see John-Steiner & Hersh, this volume). Why is the frame-breaking solution of the man in the subway not recognized as creative? By not considering the other "pole" of the magnet, we not only limit ourselves to considering solely what already exists (ethics without creativity), we also set up a future fraught with hazards born from our own myopia (creativity without ethics).

Ripple: Creativity *in* ethics

A ripple is energy flowing outward through a fluid, often in concentric circles. With this metaphor, creativity and ethics interact. Ethics is the placid but flexible fluid into which a creative contribution can be introduced. People take for granted the "calm waters" of ethical norms until someone throws a creative "stone" into them and "makes waves." Creativity is considered deviance (Becker, 1963; Stebbins, 1966, 1971): novelties that "rock the boat" are wrong and resisted, unless there is sufficient power behind their introduction to maintain momentum. Implicitly, ethics is still based on rules, and there is inertia: the "ethical waters" want to stay calm so they retain a strong, stability-oriented, defensive resistance to change.

However, ethics is fluid and has some flexibility to absorb small disturbances without upheaval. That flexibility means that rule-breaking is allowed under specific conditions. In the real world, this equates to special circumstances: It is okay to kill someone in self-defense. It is okay to tell a white lie to spare someone's feelings. It is okay to be creative in specific fields, like art. Crime and corruption are rule-breaking with malevolent intent; creativity falls into the category of rule-breaking with good intent. The intention of the creator determines the novelty's moral valence (Runco, 1993), but that intention must come with sufficient force, or at a specific angle (as with skipping stones), to have a ripple effect. The moral import of an idea or product is not recognized unless it is big enough to "make a splash," whereas most new ideas drown in indifference. People do not pay attention to the ethical entailments until the idea or product has created a wide circle of ripples.

At first, creators may be labeled rebels or troublemakers who disturb the calm (Moran, 2009a). They are rejected and sometimes resented (Monin, Sawyer, & Marquez, 2008). However, if enough cultural members "ride the wave," using the same social convergence biases (Gigerenzer, 2008) that normally tend to tamp down difference and disturbance in a culture, the ethics of the culture can converge on a new state of calm. Rather than mimicking the currently accepted "right way" (Kahneman, Slovic, & Tversky, 1989), these biases can reinforce a shifted morality (Moran, 2009a) introduced with a "transformational imperative" (Feldman, Csikszentmihalyi, & Gardner, 1994). One simulation study (Sosa & Gero, 2005) showed that even a very small number of individuals in a culture, still using the recognition heuristic (Gigerenzer, 2008) but intentionally picking the option that is *not* the most quickly recognizable, could lead to a sea change in the community. The challenge for a creator is not figuring out some complex judgment process or criteria, but keeping the idea alive long enough for it to get through most people's "familiarity screening" (see Blair & Mumford, 2007). The novelty must be not so new as to cause anxiety, but rather an adaptation that can harness the current Zeitgeist and infrastructure (Mumford & Gustafson, 1988). Creativity is other-focused and prosocial (Grant & Berry, 2011) by helping others see the not yet familiar as familiar (Moran, 2009c). Creativity impels movement, making use of the fluidity of ethics.

Smoke: Creativity *and* ethics

Smoke from a lit candle whirls and dances with the air currents, making visible the concept of turbulence. Unlike the generally regular, concentric pattern of ripples, the smoke and air create irregular patterns, chaotic flows, and agitation. When a novelty (the smoke) is introduced, it puts into motion what was already in the air; that is, the ideas that were already in the culture. The new idea can cause a cascade of changes in meaning (Bruner, 1990)—not only of facts, but sometimes of values. After the new idea is introduced, practitioners not only have to learn the new idea but also rethink their current notions.

The novelty reorganizes conceptual structures (Caughron et al., 2009; Mumford et al., 2010). Whereas ethics was considered a stabilizing property of the culture in the ripple metaphor, with the smoke metaphor ethics is no longer taken for granted as universal and stable. Ethics is a domain, just like art, science, and business are domains, and creativity can arise *in* the moral domain and change it (Gruber, 1993). Scholars using this metaphor talk of "moral relativism," not across cultures but within cultures (Haidt & Graham, 2007; Maffesoli, 1991; Wolfe, 2001). This metaphor highlights different perspectives and value systems that may be at play: not creativity imposed on or into ethics, but creativity and ethics affecting each other. The focus is the strategy for changing values and mores themselves. The dynamic is not a centripetal force to bring "deviants" into an ethical frame, but rather a turbulence of creativity and ethics interacting, often in unexpected patterns (Chonko, Wotruba, & Loe, 2003).

Scholarly work applicable to this metaphor is found more in the morality literature than the creativity literature. Some researchers started looking at "moral exemplars"—Martin Luther (Erikson, 1958), people who saved the Jews during the Second World War (Oliner & Oliner, 1988), and various others (Colby & Damon, 1992; Moran & Gardner, 2006). These moral exemplars seemed similar to exemplary creators, as described in Gardner (1993) or Csikszentmihalyi (1996), except their efforts changed *social* relations rather than *symbolic* artifacts. They changed the way people thought about each other and about social institutions.

These intentional moral creators did not heed the uniformity of a "calm waters" ethics, but rather were idealists who harnessed a particular situation (Bierly, Kolodinsky, & Charette, 2009) to bring about "creative disruption" not only in business (Vedres & Stark, 2010) but across society (Florida, 2012). These disrupters are the innovators and early adopters in diffusion of innovation studies (Rogers, 1983[1962]), the open-to-experience individuals in personality studies (Cassandro & Simonton, 2010; McElroy & Dowd, 2007), the entrepreneurs in business studies (Hall & Rosson, 2006), and the outsiders in sociological studies (Becker, 1963).

A common description of creativity in this metaphorical frame is "fruitful misalignment." The values, purposes, standards, and practices of the domain are no longer headed toward the same aim (Gardner, Csikszentmihalyi, & Damon, 2001; Moran, 2010d). Instead, the ideas from each (the "particles" in the smoke and air) are bouncing off each other. The world feels unsettled. There are still rules, but questions arise about what the rules are, what they should be, and even about the "game of the rules" (Horton & Freire, 1990; Scott, 1990). This turbulence affects both more conventional roles (such as managers; Chonko, Wotruba, & Loe, 2003) as well as other creative roles (such as entrepreneurs; Hall & Rosson, 2006).

An example is the manner in which the slave trade was finally abolished in Britain. After abolition bills failed in Parliament repeatedly over 15 years, a bill was introduced indirectly, not to abolish slavery altogether but rather to prevent the importation of slaves by British traders into territories belonging to foreign powers. The abolitionists harnessed existing laws and ethical norms in the legal field, to intersperse their "smoke" with the law's "air" and alter the meanings of existing statutes. The bill passed, which created momentum for later bills to abolish slavery completely.

Other examples include astronomers Copernicus and Galileo, who both supported a heliocentric model of the solar system and encountered tremendous opposition from religious leaders. At the time, the Church was the "air" of moral authority. Galileo was condemned to house arrest. A sun-centered system was not an astronomical issue, but a moral one about humanity's place in the universe. Both astronomers hesitated at times to publish (as did Darwin with the theory of evolution, which also affected humanity's place in the universe). Similarly, Martin Luther King, Rosa Parks, and other exemplars of the Civil Rights Movement aimed to change social relations. The 1960s were extremely turbulent times. Despite the non-violent tactics of the freedom riders and other change agents, many people were harmed or died trying to change America's moral views.

Even today, young people with aims to change the world find their creative ambitions difficult to sustain (Moran, 2010a). Especially at a life stage focused on socialization into the existing culture, there are more forces *against* these ambitions than *for* them (see A. Cropley, this volume). Especially for high school students, most of these would-be world-changers reverted to more standard "get a job and have a family" goals after two years (Moran, 2010a).

The turbulence is felt by the creators themselves as well. Trying to diversify a community's ethics can be a lonely place, as they may start out as a "minority of one" (Torrance, 1991, 1993). Furthermore, the fruits of their efforts may bring great benefits to the community in the long term, but the costs of being different are borne primarily by the creators (Putnam, 2007). A study of moral rebels shows how they endure shunning, ridicule, and other shaming techniques for doing the *right* thing (Monin, Sawyer, & Marquez, 2008). There are few social supports for diversifying or trying to change the moral domain (Haidt, Rosenberg, & Hom, 2003; Rozin et al., 1999).

Map: Creativity *of* ethics

A map spatially depicts what we know and do not know about our world. It gives us guidance where to proceed and where to avoid. Ethics codes are often considered maps for behavior, especially where the boundaries of proper behavior lie. Ethics draws the lines between "good" (known) and "bad" (unknown) lands. Creativity involves moving from the known areas of the map to the unknown. Perhaps the most fitting maps for this metaphor are the "here be dragons—beware!" maps of the Middle Ages. Still with a defensive, loss-avoidant, uncertainty-fearing foundation, this map metaphor portrays the areas ripe for creativity, but at least these opportunities are on the map! Creativity is no longer denied or shunned; it is moved offshore.

This metaphor of the creativity–ethics intersection abides mostly in venturing: seeking the "good" amid the dangerous unknown, primarily through meaning-making and exploration (Bruner, 1962, 1990), then bringing back the "bounty" to the mainland. There is a stronger sense of valuing possibilities than in the metaphors previously discussed. Ethics not only considers "what is" but could also consider "what could be," albeit cautiously.

Creativity is no longer viewed as uncontrollable in relation to a stable rule. Rather, it is a way of harnessing or controlling what could be—for those who take or support the risk of setting sail. The dragons in the map are the future, and the future is going to be encountered at some point. The belief is: perhaps we should have an agentic say in what that future will be. This perspective can be heard in mottos like Gandhi's "be the change you want to see in the world" or "invent the future."

Creators leave their cozy, familiar homes of today and seek fortunes "out there." It is a mindset shift from "small worlds" where the number of connections quickly becomes inclusive and the parameters of a problem space can be specified or assumed in advance (Granovetter, 1983; Watts & Strogatz, 1998) to "large worlds" that are not limited to the here-and-now, where uncertainty is part of the way things are and small changes can lead to big effects (Albert & Barabási, 1999). "Small worlds" with clusters and cliques help mediocre performers most, but "large worlds" open up increased possibilities for innovators to succeed as well (Guimerà et al., 2005; Uzzi & Spiro, 2005). Thus, this metaphor provides a more creativity-friendly territory by creating a bigger world to explore, at least for the brave who venture forth.

Part of the reason this metaphor depicts creativity in a more friendly light is because the upheavals occur far from mainstream society. Thus, this metaphor may exemplify Stacey's (1996) and Jaques' (1955) notions of how communities handle anxiety about uncertainty through "shadow systems" that set aside some resources for courageous explorers yet do not affect everyday life. Many institutions tend to be conservative and cordon off creativity for safety reasons into, for example, special gifted classes in education, clinical trials in medicine, or test kitchens for nouvelle cuisine.

Maps to the unknown give creators license to set aside conventional social roles or identities and avail themselves of other possibilities—just as American westward expansion provided for explorers and homesteaders. Creators discover Foucault's (1984) heterotopia, a "different place" outside of normal functioning. Then they make that place palatable to the less venturous by updating the map, removing the frightening dragons, and "filling in the gaps" with landmarks of which the less venturous can make sense.

An example from the Bible is the story of King Solomon and the dispute between two women, each claiming that the one baby is hers. Conventional moral wisdom offered to Solomon options such as he, as king, could unilaterally decide which woman got the baby, or could hold a trial. But he did something creative by using what was known (mothers love their babies and do not want them harmed) as a launchpad to venture into what was unknown (who the real mother was). He posed a threat to cut the baby in half. This gutsy move brought to the surface new meanings in formerly uncharted waters (Connell & Moran, 2008).

The map metaphor particularly highlights the need for creators to provide directions for others in the culture to understand and make use of their novelty; that is, to make their creations easier to accept. For example, creativity can direct attention to an ethical issue through symbolic means when the social structure is such that the issue cannot be approached directly without conflict or potential harm. Dorothea Lange's photographs (see Dixon & Haste, this volume) or Bono's concerts for Africa make human suffering more palatable to address because they mediate the troublesome emotions through art or music. Graphic artist Shepard Fairey's OBEY stickers and street art (see Noonan &

Gardner, this volume), as well as the Pixar movie *Up!*, call attention to the mindless acceptance of propaganda (see also John-Steiner & Hersh, this volume). These creative artifacts introduce the issues in a way that people can take in, without the message being too difficult to bear that they emotionally or intellectually shut down. Creativity develops additional possibilities for developing the mainstream by exploring terrain that others consider dangerous or off-limits.

Dough: Creativity *for* ethics

Dough combines flour, liquid, leavening, and flavorings. Different ratios of these ingredients provide a plethora of tasty results: cookies, breads, noodles, cakes, and pastry, to name a few. As a metaphor for the creativity–ethics intersection, what is important is that, once mixed, the ingredients cannot be removed. Unlike a salad, where tomatoes can easily be separated from lettuce, someone cannot separate the wheat flour from dough. The ingredients in dough have fundamentally changed each other's properties.

This metaphor differs considerably from the previous metaphors because ethics is not composed of rules. Rather, ethics represents a relationship. It is not imposing one's values on others, nor absorbing others' values. It is not a process of homogenization, but rather of embracing different perspectives. It is the perpetual coming-into-being of social relations that integrate differences. Think of cooking, where flavors and textures from a variety of ingredients contribute to a satisfying meal. This metaphor is about meaning-making chemistry: imagination embracing empathy (Johnson, 1993; Yaniv, 2012; also Narvaez & Mrkva, this volume).

This metaphor becomes all the more interesting when it embraces the "other" beyond traditional conceptions of creativity as a symbolic function and ethics as a social function. Creativity is also social— in collaboration, in interplay of ideas across minds, in judgments of value (Csikszentmihalyi, 1988; John-Steiner, 1997, 2000). Ethics is also symbolic—in rituals and documents and gestures (see Dixon & Haste, this volume). Creativity and ethics are both dimensions applicable to *every* situation and domain. What seems like only a "personal" choice not a "moral" decision (Turiel, 1983) may be a case of myopia. Overeating or smoking, for example, are personal lifestyle choices in the short term, but they have considerable ethical ramifications regarding public health, healthcare costs, and use of common resources like bus seats or road maintenance in the longer term.

Although only a few researchers consider this metaphorical foundation (Grant & Berry, 2011; Muhr, 2010), several practicing artists and
social activists support it. The notion of embracing is often seen in
"dialogue" approaches to engagement of diversity, difference, or the
new (such as thedialogueproject.org; Calabria et al., 2008; D'Arlach,
Sanchez, & Feuer, 2009) as well as in perspectives on mainstream culture from non-mainstream individuals (Scott, 1990). Theater of the
Oppressed, for example, uses audience members as "spect-actors" to
explore, analyze, and transform social reality (Boal, 1993). Crossroads
Charlotte in North Carolina is a program that asks citizens to share stories depicting plausible futures for the city (www.crossroadscharlotte.
org). Saffron in Chicago produced an original play written by teens,
inspired by real-life events of immigrants and the working class, to
provide perspectives on issues of equality and opportunity (Metz, 2005).

How might this metaphorical understanding of creativity–ethics play
out in a situation? In the Bible, the story of Jesus and the Pharisees
addressing the fate of an adulterous woman may be an example
(Connell & Moran, 2008). Jesus says, "Let he who is without sin cast
the first stone." This statement, which forced the Pharisees into an
exercise in perspective-taking, shifted the relational fabric of the situation. The "other" (the woman) disrupted the Pharisees' belief in their
self-righteousness. Once that shift has been made—once the two perspectives are mixed—the self-righteous perspective could not be "pulled
back out" intact.

Another example comes from constitutional history. Constitutions
are not rules per se, but rather they address how to make the rules of a
society. They concern the wider dynamics of rulings, rather than rules-
as-given and rule-breaking (Cua, 1978; Havel, 1997). The Constitution
Museum in Philadelphia, Pennsylvania, tells the history of the development of the US Constitution as legislators over time interpreted it
to make rules for changing the meaning of a "person" to include, for
example, non-white races and women. More recent interpretations suggest that corporations, animals, and the whole planet are also "persons."
Constitutions guide who or what is embraced that, afterward, would be
difficult to unrecognize.

This metaphor highlights that creativity and ethics, as they interact,
change the properties of each other. Rather than taking snapshots of
creativity and ethics in a given situation, this metaphor emphasizes
the dynamics of change itself. It provides a two-way "zone of proximal
development" (Vygotsky, 1978) in which creativity and ethics both are
active participants (Moran, 2010b).

What *should* we pursue?

This metaphor-based introduction aims to kindle interest and generate ideas for further investigation. As you read the chapters that follow, consider how their arguments and findings may build on one or more of these metaphors. Or consider how these metaphors may not be independent types; there may be some relationship or ordinality.

One ranking criterion may be *time*. Rule-based creativity *or* ethics (magnet) removes time from the equation. The rules are timeless. Intent-based creativity *in* ethics (ripple) describes a one-point-in-time crossing of two domains. The focus is on the moment the novelty is instigated in a culture. Strategy-based creativity *and* ethics (smoke) also focuses on the moment of instigation, but orients more toward how the instigation unfolds. Mostly, the progression is expected to be unstable. Venture-focused creativity *of* ethics (map) posits the future as most important. The destination, even if unknown, supersedes the departure point as a focus. Embrace-focused creativity *for* ethics (dough) makes the movement of time preeminent. Time shows the process of change—how interactions among entities alter those entities. It is not any particular point in time—the past and tradition as in rules, nor the present as in intent, nor the future as in ventures—that matters, but a rethinking that time *flows*, and both creativity and ethics affect the flow of culture.

Twenty years ago, psychologist Howard Gruber (1993) called the creativity–ethics connection an imperative: "ought implies can implies create." Although "the tragedy of the commons" (Hardin, 1968) is usually conceived as overusing existing public resources—taking *from* a common good that may not be replaceable—the true tragedy of the commons may involve not allowing sufficient contribution *to* the common good, not providing the means for the commons to grow. Out of fear or lack of foresight, perhaps we underutilize or misuse one of the most powerful resources for cultural progress: the mind's flexibility to make and remake ideas, meanings, and the opportunities they entail. Using this powerful creative resource ethically requires a contribution mindset: we need to focus not on what we get out of the commons but what we put into it. We need to look around and think ahead regarding how our actions (or inactions) contribute to how the future unfolds not only for ourselves, but for others and future generations. We need multiple perspectives because no one point of view has the experience, capabilities, or imagination to elaborate all possible consequences. Although interacting with others different from ourselves—especially those with different moral foundations (Haidt,

Koller, & Dias, 1993; Haidt, Rosenberg, & Hom, 2003)—may be more difficult than associating with like-minded people, such collaborations of diversity are imperative to stretching ourselves individually and communally in positive directions (Moran & John-Steiner, 2004) and to maintaining humility as we aim to change the world for the better (Moran, 2012).

It is important for us to be more intentional about creativity, how it works, and whether its effects are something we *should* strive for (Tenner, 1996; Thackara, 2006). The ethicality aspect of creativity has been relatively ignored until recently, and what scholarly discussion has occurred is mostly based on assumptions and far from convergent. One view is that creativity is prima facie amoral (Runco, 2010): creativity itself is neither right nor wrong, and creators are not responsible for their creations or those creations' consequences. A novelty could be an error or eccentric, but it does not necessarily carry ethical entailments just by being new. The magnet metaphor represents this stance. Another view is that creativity is prima facie unethical because it creates social and cultural turbulence (Hall & Rosson, 2006), muddles ethical lines (Gino & Ariely, 2012; Vincent & Goncalo, this volume), and may cause a "moral spillover effect" (Mullen & Nadler, 2008) whereby an increasing number of people become unmoored from conventional behavioral controls (Gino, Ayal, & Ariely, 2009). The ripple and smoke metaphors depict this creativity-as-unethical perspective, although the fear of creativity is less with the smoke metaphor than the ripple metaphor. Still another view is that creativity is prima facie ethical because it brings into being ideas, products, and tools that increase our individual and collective capacities (Bruner, 1962; Guilford, 1950; see Richards, this volume). Creativity improves on what already exists (Moran, 2010e). The map metaphor depicts this perspective.

An increasing number of scholars are challenging these a priori stances toward the creativity–ethics intersection and instead suggesting that people should be more mindful and sensitive to nuances in the relationship between what is new and what is good (Csikszentmihalyi & Nakamura, 2007; Gruber, 1993; Moran, 2010d; Runco & Nemiro, 2003). The dough metaphor represents this stance. Implicitly, this volume endorses this final stance: rather than stating that creativity is amoral, ethical, or unethical, let us consider by what means and under what conditions creativity and ethics might manifest each of these metaphorical interactions.

Creativity spawns uncertainty, or at least makes uncertainty more salient (Hall & Rosson, 2006; Kagan, 2009). This uncertainty manifests not only in how a creative idea or product actually will function in the real world (see D. Cropley and Finn, this volume), but also how people might adapt to it, create new venues or uses for it, and proliferate its impact beyond its initial intentioned applications (Dudley, 2010; Tenner, 1996). Introducing something creative not only introduces the idea or product, but also can launch a system that can take on a life of its own (Hall & Rosson, 2006; Jasper, 2010; Lloyd, 2009; Thackara, 2006). What will happen—what *could* happen—then? We are unclear how to judge truly novel ideas and products (Licuanan, Dailey, & Mumford, 2007). For decades, creativity research has explored the criteria we may use to judge the novelty of the product compared to what is "already out there" (Amabile, 1983), the functionality of the product to work as expected (see D. Cropley, this volume), and the usefulness of the product to various constituencies (see Mumford et al., this volume). In this volume, some authors ask creators to envision criteria regarding the morality and ethics of their products—their benefits and harms to others (Haste, 1993; see Dixon & Haste, John-Steiner & Hersh, Li & Csikszentmihalyi, this volume). How should those judgments proceed? Other authors ask how we might incorporate creativity into ethical codes. Case studies and histories demonstrate that ethics is partially driven by creative approaches, such as moral and intellectual exemplars who overturned old social patterns (Colby & Damon, 1992; Daloz et al., 1996; Gardner, 1993; Gruber, 1981; Noonan & Gruber and John-Steiner & Hersh, this volume). How can we keep ethics supple? As one reviewer described the movie *Amazing Grace* and its story of how British aristocrat William Wilberforce and his colleagues rallied for the end of the slave trade: ethical creativity and creative ethics "bare our blind spots [and] set a template for what follows" (Powers, 2007).

References

Albert, R., & Barabási, A.L. (1999). Emergence of scaling in random networks. *Science, 286*(5439), 509–512.

Amabile, T. M. (1983). The social psychology of creativity: A componential conceptualization. *Personality Processes and Individual Differences, 45*(2), 357–376.

Becker, H. S. (1963). *Outsiders: Studies in the Sociology of Deviance.* New York: Free Press.

Bierly, P. E. III, Kolodinsky, R. W., & Charette, B. J. (2009). Understanding the complex relationship between creativity and ethical ideologies. *Journal of Business Ethics, 86*, 101–112.

Blair, C. S., & Mumford, M. D. (2007). Errors in idea evaluation: Preference for the unoriginal? *Journal of Creative Behavior, 41*(3), 197–222.

Boal, A. (1993). *Theater of the Oppressed*. New York: Theatre Communications Group.

Bruner, J. S. (1962). The conditions of creativity. In J. S. Bruner, *On Knowing: Essays for the Left Hand*. Cambridge: Cambridge University Press.

Bruner, J. S. (1990). *Acts of Meaning*. Cambridge, MA: Harvard University Press.

Butcher, S. I. (2005). *The Origins of the Russell-Einstein Manifesto*. Washington, DC: Pugwash Conferences on Science and World Affairs. http://www.pugwash.org/ publication/phs/history9.pdf, accessed July 8, 2013.

Calabria, M., Lentz, R., Karecki, M., Omar, I., & Cusato, M. (2008). *Daring to Embrace the Other: Franciscans and Muslims in Dialogue*. Bonaventure, NY: Franciscan Institute.

Cassandro, V. J., & Simonton, D. K. (2010). Versatility, openness to experience, and topical diversity in creative products: An exploratory historiometric analysis of scientists, philosophers, and writers. *Journal of Creative Behavior, 44*(1), 1–18.

Caughron, J. J., Shipman, A. S., Beeler, C. K., & Mumford, M. D. (2009). Social innovation: Thinking about changing the system. *International Journal of Creativity and Problem Solving, 19*(1), 7–32.

Chonko, L. B., Wotruba, T. R., & Loe, T. W. (2003). Ethics code familiarity and usefulness: Views on idealist and relativist managers under varying conditions of turbulence. *Journal of Business Ethics, 42*(3), 237–252.

Colby, A., & Damon, W. (1992). *Some Do Care: Contemporary Lives of Moral Commitment*. New York: Free Press.

Connell, M. W., & Moran, S. (2008). All the wiser: Wisdom from a systems perspective. Invited talk at University of Chicago Arete Initiative, Chicago, June.

Csikszentmihalyi, M. (1988). Society, culture, and person: A systems view of creativity. In R. J. Sternberg (ed.), *The Nature of Creativity* (pp. 325–339). New York: Cambridge University Press.

Csikszentmihalyi, M. (1996). *Creativity*. New York: HarperCollins.

Csikszentmihalyi, M. (1999). Implications of a systems perspective for the study of creativity. In R. J. Sternberg (ed.), *Handbook of Creativity* (pp. 313–338). Cambridge: Cambridge University Press.

Csikszentmihalyi, M., & Nakamura, J. (2007). Creativity and responsibility. In H. Gardner (ed.), *Responsibility at Work* (pp. 64–80). San Francisco, CA: Jossey-Bass.

Cua, A. S. (1978). *Dimensions of Moral Creativity: Paradigms, Principles, and Ideals*. University Park: Pennsylvania State University Press.

D'Arlach, L., Sanchez, B., & Feuer, R. (2009). Voices from the community: A case for reciprocity in service-learning. *Michigan Journal of Community Service Learning, 16*(1), 5–16.

Daloz, L., Keen, C., Keen, J., & Parks, S. (1996). *Common Fire: Lives of Commitment in a Complex World*. Boston, MA: Beacon Press.

Debatin, B., Lovejoy, J. P., Horn, A.-K., & Hughes, B. N. (2009). Facebook and online privacy: Attitudes, behaviors and unintended consequences. *Journal of Computer Mediated Communication, 15*(1), 83–108.

Dudley, W. C. (2010). Asset bubble and the implications for central bank policy. Speech at Economic Club of New York, New York, April 7. http://www.bis.org/ review/r100409c.pdf, accessed November 11, 2013.

Erikson, E. (1958). *Young Man Luther*. New York: W. W. Norton.

Feldman, D. H., Csikszentmihalyi, M., & Gardner, H. (1994). *Changing the World: A Framework for the Study of Creativity*. Westport, CT: Praeger.

Florida, R. L. (2012). *The Rise of the Creative Class: Revisited*. New York: Basic Books.

Foucault, M. (1984). Of other spaces, heterotopias. *Architecture, Mouvement, Continuité, 5*, 46–49.

Friedan, B. (1963). *The Feminine Mystique*. New York: W. W. Norton.

Gardner, H. (1993). *Creating Minds*. New York: Basic Books.

Gardner, H., Csikszentmihalyi, M., & Damon, W. (2001). *Good Work: When Ethics and Excellence Meet*. New York: Basic Books.

Gigerenzer, G. (2008). Why heuristics work. *Perspectives on Psychological Science, 3*(1), 20–29.

Gigerenzer, G. (2010). Moral satisficing: Rethinking moral behavior as bounded rationality. *Topics in Cognitive Science, 2*(3), 528–554.

Gino, F., & Ariely, D. (2012). The dark side of creativity: Original thinkers can be more dishonest. *Journal of Personality and Social Psychology, 102*(3), 445–459.

Gino, F., Ayal, S., & Ariely, D. (2009). Contagion and differentiation in unethical behavior: The effect of one bad apple on the barrel. *Psychological Science, 20*(3): 393–398.

Glaveanu, V. P. (2011). Creativity as cultural participation. *Journal for the Theory of Social Behaviour, 41*(1), 48–67.

Granovetter, M. (1983). The strength of weak ties: A network theory revisited. *Sociological Theory, 1*, 201–233.

Grant, A. M., & Berry, J. W. (2011). The necessity of others is the mother of invention: Intrinsic and prosocial motivations, perspective taking, and creativity. *Academy of Management Journal, 54*(1), 73–96.

Gruber, H. E. (1981). *Darwin on Man: A Psychological Study of Scientific Creativity*. Chicago, IL: University of Chicago Press.

Gruber, H. E. (1993). Creativity in the moral domain: Ought implies can implies create. *Creativity Research Journal, 6*(1–2), 3–15.

Guilford, J. P. (1950). Creativity. *American Psychologist, 5*, 444–454.

Guimerà, R., Uzzi, B., Spiro, J., & Amaral, J. A. N. (2005). Team assembly mechanisms determine collaboration network structure and team performance. *Science, 308*(5722), 697–702.

Haidt, J, & Graham, J. (2007). When morality opposes justice: Conservatives have moral intuitions that liberals may not recognize. *Social Justice Research, 20*, 98–116.

Haidt, J., Koller, S. H., & Dias, M. G. (1993). Affect, culture, and morality, or is it wrong to eat your dog? *Journal of Personality and Social Psychology, 65*(4), 613–328.

Haidt, J., Rosenberg, E., & Hom, H. (2003). Differentiating diversities: Moral diversity is not like other kinds. *Journal of Applied Social Psychology, 33*(1), 1–36.

Hall, J., & Rosson, P. (2006). The impact of technological turbulence on entrepreneurial behavior, social norms and ethics: Three internet-based cases. *Journal of Business Ethics, 64*, 231–248.

Hardin, G. (1968). The tragedy of the commons. *Science, 162*, 1243–1248.

Haste, H. (1993). Moral creativity and education for citizenship. *Creativity Research Journal, 6*, 153–164.

Havel, V. (1997). *The Art of the Impossible: Politics as Morality in Practice.* New York: Knopf.
Hersh, R. (1990). Mathematics and ethics. *Mathematical Intelligencer, 12,* 13–15.
Hirsh, J. B., Mar, R. A., & Peterson, J. B. (2012). Psychological entropy: A framework for understanding uncertainty-related anxiety. *Psychological Review, 119*(2), 304–310.
Horton, M., & Freire, P. (1990). *We Make the Road by Walking: Conversations on Education and Social Change* (B. Bell, J. Gaventa, & J. Peters, eds.). Philadelphia, PA: Temple University Press.
Jaques, E. (1955). Social systems as a defense against persecutory and depressive anxiety. In M. Klein, P. Heimann, & R.E. Money-Kyrle (eds.), *New Directions in Psychoanalysis* (pp. 478–498). London: Tavistock.
Jasper, J. M. (2010). The innovation dilemma: Some risks of creativity in strategic agency. In D. H. Cropley, A. J. Cropley, J. C. Kaufman, & M. A. Runco (eds.), *The Dark Side of Creativity* (pp. 91–113). New York: Cambridge University Press.
John-Steiner, V. (1997). *Notebooks of the Mind: Explorations in Thinking,* rev. edn. New York: Oxford University Press.
John-Steiner, V. (2000). *Creative Collaboration.* New York: Oxford University Press.
Johnson, M. (1993). *Moral Imagination: Implications of Cognitive Science for Ethics.* Chicago, IL: University of Chicago Press.
Kagan, J. (2009). Categories of novelty and states of uncertainty. *Review of General Psychology, 13*(4), 290–301.
Kahneman, D., Slovic, P., & Tversky, A. (1989). *Judgment under Uncertainty: Heuristics and Biases.* New York: Cambridge University Press.
Licuanan, B. F., Dailey, L. R., & Mumford, M. D. (2007). Idea evaluation: Error in evaluating highly original ideas. *Journal of Creative Behavior, 41*(1), 1–27.
Lloyd, P. (2009). Ethical imagination and design. *Design Studies, 30,* 154–168.
Maffesoli, M. (1991). The ethic of aesthetics. *Theory, Culture and Society, 8*(1), 7–20.
McElroy, T., & Dowd, K. (2007). Susceptibility to anchoring effects: How openness-to-experience influences responses to anchoring cues. *Judgment and Decision Making, 1,* 48–53.
Metz, N. (2005). "Saffron" gives us taste of immigrants' lives in Chicago. *Chicago Tribune,* July 5. http://articles.chicagotribune.com/2005-07-05/features/0507040197_1_restaurant-immigrants-meal, accessed November 11, 2013.
Monin, B., Sawyer, P. J., & Marquez, M. J. (2008). The rejection of moral rebels: Resenting those who do the right thing. *Journal of Personality and Social Psychology, 95*(1), 76–93.
Moran, S. (2009a). Creativity: A systems perspective. In T. Richards, M. Runco, & S. Moger (eds.), *The Routledge Companion to Creativity* (pp. 292–301). London: Routledge.
Moran, S. (2009b). Metaphor foundations of creativity research: Boundary versus organism. *Journal of Creative Behavior, 43*(1), 1–22.
Moran, S. (2009c). What role does commitment play among writers with different levels of creativity? *Creativity Research Journal, 21*(2–3), 243–257.
Moran, S. (2010a). Changing the world: Tolerance and creativity aspirations among American youth. *High Ability Studies, 21*(2), 117–132.
Moran, S. (2010b). Commitment and creativity: Transforming experience into art. In C. Connery, V. John-Steiner, & A. Marjanovic-Shane (eds.), *Vygotsky and*

Creativity: A Cultural-Historical Approach to Meaning-Making, Play, and the Arts
(pp. 141–160). New York: Peter Lang.
Moran, S. (2010c). Creativity in school. In K. S. Littleton, C. Wood, &
J. K. Staarman (eds.), *International Handbook of Psychology in Education*
(pp. 319–360). Bingley: Emerald Group.
Moran, S. (2010d). Returning to the Good Work Project's roots: Can creative work
be humane? In H. Gardner (ed.), *Good Work: Theory and Practice* (pp. 127–152).
Cambridge, MA: Good Work Project.
Moran, S. (2010e). The roles of creativity in society. In J. C. Kaufman &
R. J. Sternberg (eds.), *The Cambridge Handbook of Creativity* (pp. 74–90).
New York: Cambridge University Press.
Moran, S. (2012). Book review: The dark side of creativity (D. H. Cropley, A. J.
Cropley, J. C. Kaufman, & M.A. Runco, eds.). *Psychology of Aesthetics, Creativity,
and the Arts, 6*(3), 295–296.
Moran, S., & Gardner, H. (2006). Extraordinary cognitive achievements: A devel-
opmental and systems analysis. In W. Damon (series ed.) & D. Kuhn &
R. S. Siegler (vol. eds.), *Handbook of Child Psychology, Vol. 2: Cognition, Perception,
and Language*, 6th edn. (pp. 905–949). New York: John Wiley & Sons.
Moran, S., & John-Steiner, V. (2003). Creativity in the making: Vygotsky's
contribution to the dialectic of creativity and development. In K. Sawyer,
V. John-Steiner, S. Moran, R. Sternberg, D. Feldman, et al. (eds.), *Creativity and
Development* (pp. 61–90). New York: Oxford University Press.
Moran, S., & John-Steiner, V. (2004). How collaboration in creative work
impacts identity and motivation. In D. Miell & K. Littleton (eds.), *Collabora-
tive Creativity: Contemporary Perspectives* (pp. 11–25). London: Free Association
Books.
Mueller, S., Melwani, S., & Goncalo, J. A. (2012). The bias against creativity: Why
people desire but reject creative ideas. *Psychological Science, 23*, 13–17.
Muhr, S. L. (2010). Ethical interruption and the creative process: A reflection on
the new. *Culture and Organization, 16*(1), 73–86.
Mullen, E., & Nadler, J. (2008). Moral spillovers; The effect of moral violations on
deviant behavior. *Journal of Experimental Social Psychology, 44*, 1239–1245.
Mumford, M. D., & Gustafson, S. B. (1988). Creativity syndrome: Integration,
application, and innovation. *Psychological Bulletin, 103*(1), 27–43.
Mumford, M. D., Waples, E. P., Antes, A. L., Brown, R. P., Connelly, S., et al. (2010).
Creativity and ethics: The relationship of creative and ethical problem-solving.
Creativity Research Journal, 22(1), 74–89.
Oliner, S. P., & Oliner, P. M. (1988). *The Altruistic Personality: Rescuers of Jews in
Nazi Europe*. New York: Free Press.
Powers, J. (2007). "Amazing Grace" tells the story of British abolition. NPR
Fresh Air from WHYY, March 15. http://www.npr.org/templates/story/story.
php?storyId=8926395, accessed November 11, 2013.
Putnam, R. D. (2007). E pluribus unum: Diversity and community in the twenty-
first century (the 2006 Johan Skytte Prize Lecture). *Scandinavian Political Studies,
30*, 137–174.
Rogers, E. M. (1983[1962]). *Diffusion of Innovations*, 3rd edn. New York: Free Press.
Rozin, P., Lowery, L., Imada, S., & Haidt, J. (1999). The CAD triad hypothesis:
A mapping between three moral emotions (contempt, anger, disgust) and three

moral codes (community, autonomy, divinity). *Journal of Personality and Social Psychology, 76,* 574–586.

Runco, M. A. (1993). Creative morality: Intentional and unconventional. *Creativity Research Journal, 6*(1–2), 17–28.

Runco, M. A. (2010). Creativity has no dark side. In D. H. Cropley, A. J. Cropley, J. C. Kaufman, & M. A. Runco (eds.), *The Dark Side of Creativity* (pp. 15–32). New York: Cambridge University Press.

Runco, M. A., & Jaeger, G. J. (2012). The standard definition of creativity. *Creativity Research Journal, 24*(1), 92–96.

Runco, M. A., & Nemiro, J. (2003). Creativity in the moral domain: Integration and implications. *Creativity Research Journal, 15*(1), 91–105.

Sandel, M. J. (2007). *The Case against Perfection: Ethics in the Age of Genetic Engineering.* Cambridge, MA: Harvard University Press.

Scott, J. C. (1990). *Domination and the Arts of Resistance: Hidden Transcripts.* New Haven, CT: Yale University Press.

Sosa, R. (2011). Understanding the future of change agency in sustainability through cellular automata scenarios: The role of timing. *Sustainability, 3,* 578–595.

Sosa, R. S., & Gero, J. S. (2004). Diffusion of creative design: Gatekeeping effects. *International Journal of Design Computing, 4,* 1–10.

Sosa, R. S., & Gero, J. S. (2005). A computational study of creativity in design: The role of society. *Artificial Intelligence for Engineering Design, Analysis and Manufacturing, 19*(4), 229–244.

Stacey, R. D. (1996). *Complexity and Creativity in Organizations.* San Francisco, CA: Berrett-Koehler.

Stebbins, R. A. (1966). Class, status, and power among jazz and commercial musicians. *Sociological Quarterly, 7*(1), 197–213.

Stebbins, R. A. (1971). *Commitment to Deviance.* Westport, CT: Greenwood Publishing.

Stein, M. I. (1993). Moral issues facing intermediaries between creators and the public. *Creativity Research Journal, 6,* 197–200.

Subotnik, R. F., Jarvin, L., Moga, E., & Sternberg, R. J. (2003). Wisdom from gatekeepers: Secrets of success in music performance. *Bulletin of Psychology and the Arts, 4*(1), 5–9.

Tenner, E. (1996). *Why Things Bite Back: Technology and the Revenge of Unintended Consequences.* New York: Vintage.

Thackara, J. (2006). *In the Bubble: Designing in a Complex World.* Cambridge, MA: MIT Press.

Torrance, E. P. (1991). The beyonders and their characteristics. *Creative Child and Adult Quarterly, 16,* 69–79.

Torrance, E. P. (1993). The beyonders in a thirty year longitudinal study of creative achievement. *Roeper Review, 15,* 131–134.

Turiel, E. (1983). *The Development of Social Knowledge: Morality and Convention.* New York: Cambridge University Press.

Uzzi, B., & Spiro, J. (2005). Collaboration and creativity: The small world problem. *American Journal of Sociology, 111*(2), 447–504.

Valsiner, J. (2000). *Culture and Human Development.* Thousand Oaks, CA: Sage.

Vedres, B., & Stark, D. (2010). Structural folds: Generative disruption in overlapping groups. *American Journal of Sociology, 115*(4), 1150–1190.

Vygotsky, L. S. (1978). *Mind in Society: The Development of Higher Psychological Processes* (M. Cole, V. John-Steiner, S. Scribner, & E. Souberman, eds.). Cambridge, MA: Harvard University Press.

Watts, D. J., & Strogatz, S. H. (1998). Collective dynamics of "small-world" networks. *Nature, 393*(6684): 440–442.

Wolfe, A. (2001). *Moral Freedom.* New York: W. W. Norton.

Yaniv, D. (2012). Dynamics of creativity and empathy in role reversal: Contributions from neuroscience. *Review of General Psychology, 16*(1), 70–77.

Part I

What Are the Moral Mental Mechanisms Involved in Creativity, and How Do They Develop?

The following five chapters examine how mental mechanisms—such as imagination, embodied knowledge, reasoning, integrity, and discernment—influence how morality and ethics play a part in creative work.

1

The Development of Moral Imagination

Darcia Narvaez and Kellen Mrkva
University of Notre Dame, USA and University of Colorado Boulder, USA

> *A man, to be greatly good, must imagine intensely and comprehensively; he must put himself in the place of many others... the great instrument of moral good is the imagination.*
> Percy Bysshe Shelley, 1821, p. 13

Creativity has been defined as the ability to generate ideas that are original and unexpected, but are considered useful or important (Sternberg, 1999). Moral imagination involves not only the ability to generate useful ideas, but also the abilities to form ideas about what is good and right, and to put the best ideas into action for the service of others. This involves sensitivity to the people and lifescapes at hand. The everyday world is populated with opportunities to steer consciously through the shoals of social relationships and decide what sort of agent to be. Research into mental preoccupations indicates that individuals ponder moral and relational issues much of the time (Klinger, 1978). Thus, on a daily basis, people employ one of humanity's greatest gifts: moral imagination. But what fosters the development of moral imagination and determines to what extent it is used to benefit humanity? How does the morally imaginative individual utilize emotional and social experiences, reasoning, and selection to produce imaginative moral action? These are the questions that this chapter addresses.

The interest in psychological research on morality is growing rapidly (Haidt, 2007) and spreading to a large number of fields. Yet, it is rare to encounter a moral psychology study that examines creativity, or theorists who give much room for creativity in their accounts of moral functioning. Although there is at least some empirical research that will shed light on these topics, John Dewey's philosophical accounts may provide the greatest insights.

Dewey's conceptions of moral imagination perhaps best advanced understanding of the relationship between creativity and morality (Fesmire, 2003). He conceived of imagination as a dramatic rehearsal in which people creatively explore and rehearse alternative courses of actions such that likely outcomes and impacts on others would guide moral decisions. The moral life involves co-authoring the future with others through dialogue and feedback on imagined alternatives, but also developing keen perception and flexible responses to each situation.

The place of moral imagination

The prevailing view on the interaction between creativity, deliberation, and morality is captured by deontological philosophy. This perspective emphasizes moral deliberation as conscious reasoning, which is assumed to exist apart from emotion (Kant, 1949). Emotions are considered to be inconsistent, unreliable, and irrational, and, thus, to be avoided. A deontological approach has little room for moral imagination. Imagination was considered to be in the realm of aesthetics and outside of morality (Johnson, 1993). The situations typically discussed are those with clear rules. Kohlberg's measures pitted values against one another and scored responses within certain established boundaries (see Gibbs, 2003). The roles of creative and practical thinking, and the influence of emotions and situational considerations, were downplayed, if not considered completely irrelevant to the goal of measuring moral reasoning capacities (Fesmire, 2003).

However, explicit reasoning is insufficient for the moral life. From a neurobiological perspective, the emphasis on conscious reasoning and selection of principles is dominated by the intellectual "left brain" (McGilchrist, 2009). The intellect typically comprises the conscious aspects of the mind, which tend to minimize the vast tacit knowledge of, and behavioral control by, the rest of the brain. Relying on intellect alone signals that the intuitive mind and emotional intelligence are underutilized or underdeveloped (Narvaez, 2014).

In contrast to a heavy emphasis on the application of reason to moral decisions and judgments, other philosophers emphasize emotion as the source of moral judgments. Moral judgments spring forth without effort or worry. Reason is used only to defend the intuitive response. Building on Hume's (1969[1739]) view, Haidt (2001) proposes a social intuitionist theory that emphasizes instantaneous moral judgment, defined as evaluations of other people's actions and characters. As with Kohlberg's studies the prototypical situations and methods deployed by Haidt and others do not enlist creativity. The unusual and emotionally strident situations create a quick positive or negative response that biases the conclusions these authors make about reason, creativity, and morality (Monin, Pizarro, & Beer, 2007).

A third perspective directly addresses creativity in moral situations. Building on social intuitionism's view that reasoning is used for post hoc rationalization, Ariely and colleagues conclude that creativity may actually *increase* unethical behavior (Gino & Ariely, 2012). This is because creativity makes individuals better at inventing justifications for cheating and more skilled at defending personal moral goodness after moral violations. In one study, priming individuals with creative words led to more cheating, suggesting a causal link between creativity and cheating. Despite these concerns, like John Dewey (2009[1908]), Mark Johnson (1993), and others, we believe that moral imagination contributes positively to moral functioning in most circumstances. Moral imagination relies on different types of intelligence: cognitive, social, and emotional. Emotional intelligence may be foundational for the others.

Emotion and moral development

When emotion systems are misdeveloped, morality can go awry. Early life shapes the emotional and cognitive capabilities that underlie morality and imagination (Greenspan & Shanker, 2004). A child is born with only one quarter of his or her brain developed; caregivers co-construct 75 percent of the brain of a full-term infant in the first years after birth (Trevathan, 2011). As a dynamic system, early life experience on multiple levels sets the stage for the rest of life. Caregivers shape the thresholds for numerous brain/body circuitries, and much of this involves the neuroendocrine and emotion systems (Meaney, 2010; Schore, 2003a, b). Too much stress at the wrong times in the first years of life can foster a stress-reactive brain, setting up a self-protective personality (Narvaez, 2008, 2014). For example, when infants do not receive physical comforting in timely ways, the vagus nerve can be mistuned, leading to long-term difficulties with social relations as well as numerous health problems (Narvaez, in press; Porges, 2011). When an infant is distressed too much during gestation or postnatal life, the hypothalamic-pituitary-adrenal axis (HPA) can be tuned to be hyper- or hypo-active (Lupien et al., 2009).

Stress affects imagination and creativity. Children who suffer from posttraumatic stress disorder have difficulty with daydreaming and symbolic play, which consolidate meaning, affect, and representation (Reid, 1999; Slade, 1987, 1994). Whenever the stress response is active, it draws energy away from higher-order thinking capacities, influencing how and how well a person imagines and relates to others (Sapolsky, 2004). If stress and trauma occur early in life, neurobiological systems never reach their optimal trajectories (Shonkoff et al., 2012). Well-rehearsed stress states become traits. In this way, neurobiological systems influence morality, setting up propensities to use different social and moral mindsets (Narvaez, 2014). What the brain's capacities look like has much to do with early life experience, when brain system connections are being established.

Many philosophical traditions and psychological theories have underestimated the role emotion plays in moral functioning, although there have been some exceptions (for example, Hume). Even among those who emphasize emotion descriptively, emotion has been viewed overwhelmingly as normatively disruptive and as impairing moral judgment (Ben Ze'ev, 2000). Philosophers have viewed emotion as passive, undependable, and even primitive and bestial. In reality, individuals use their emotional experience to think in inclusive and integrative ways (Isen & Daubman, 1984), build social relationships, and broaden creative possibilities (Fredrickson & Branigan, 2005).

In recent decades, it has become clear that emotions serve as informatory guides to adaptive judgments and behavior (Panksepp, 1998; Slovic et al., 2002). They serve as cues to value, the relevance of stimuli, and whether our actions are successful (Panksepp, 1998). In the moral domain specifically, emotions reflect our goals and values and help us respond flexibly and adaptively (Pizarro, 2000). They can be used imaginatively in attending to the morally relevant aspects of a situation, selecting moral goals, integrating values, and being sensitive to other individuals (see "Everyday Moral Imagination," later in this chapter). They form the substrate for moral motivation and action (Blasi, 1999). Triune ethics theory takes this perspective.

Extending Dewey's moral imagination

Triune ethics theory (TET; Narvaez, 2008) asserts that humans rely on a variety of neurobiologically derived moral mindsets resulting from evolved global brain states (MacLean, 1990). Individuals habitually can favor one mindset or fluctuate among several. Before delineating the theory in detail, it is instructive to note the tremendous overlap and agreement between the conceptions of moral imagination and moral functioning of Dewey and TET.

First, both theories emphasize that moral imagination requires being sensitive to the morally relevant aspects of a situation, envisioning different alternatives for action, and thinking about the ramifications of a particular action for the people involved (Fesmire, 2003; Narvaez & Vaydich, 2008; Somerville, 2006). These capacities rely on finely tuned perception, which is strongly affected by where one habitually places one's attention (Murdoch, 1989). If one's attention is captivated by perceived threat cues, then moral perception will be narrowed to what is self-protective.

Second, both Dewey and TET emphasize the social nature of moral imagination and the need for flexible, open thinking. Our studies examining TET orientations have found strong correlations between moral imagination and both openness to experience and agreeableness. Both theories also posit that flexible thinking and the ability to adapt in ongoing social relationships characterize imaginative behavior. Dewey asserts that individuals with flexibility and the ability to deal with ambiguity in imaginative ways are better able to perceive moral situations and act effectively. Dewey views moral behavior as co-authored with others and as occurring in uncertain or ambiguous situations.

Third, both TET and Dewey emphasize that moral imagination involves self-regulation, or an ability to put beliefs and goals into action. Dewey idealizes the person who is able to regulate behavior based on imagined effects and to inspect beliefs for their value in action. TET goes so far as to posit neurobiological roots of moral imagination in which individuals engage the prefrontal cortex to self-regulate, prevent harmful behaviors through "free won't," and engage in reflective abstraction.

Fourth, both TET and Dewey emphasize the importance of harmony in dealing with multiple values, such as autonomy and community. TET asserts the importance of coordinating emotion and reason, the conscious mind with the adaptive unconscious, and goals with mood and energy. Combining the insights of Dewey and TET creates a better understanding of the processes in which moral exemplars engage when facing a moral situation. Individuals can take advantage of the power and intelligence of the moral emotions while using regulation and metacognition to ensure that they guide their behavior toward fulfillment of moral goals and virtues.

Triune ethics

Thus, triune ethics theory extends Dewey's keen insights while delineating the developmental and neurobiological substrates of moral imagination. It develops a view of various moral mindsets on which people rely, which are formed during reciprocal interactions with caregivers in early life and other sensitive periods in life. The list of types is in Table 1.1.

There are emergent rules about getting along with others that develop in supportive environments and foster right-brain development (Schore, 1994, 2003a, b). Optimal early life offers the experience of reciprocal interaction through intersubjectivity and mutual influence. Intersubjective responsivity—attending to and responding to social signaling in a collaborative manner—is a creative response. Babies are ready for playful, creative protonarrative co-construction with caregivers at birth (Trevarthen, 2005). Baby and caregiver create their own stories through reciprocal, sensitive communication. This type of "companionship care" fosters three types of attachment (Narvaez, 2014). *Protective attachment* is like imprinting, a desire for physical proximity, and is evident even in abused children. *Warmth attachment* is emotional connection to the caregiver, which facilitates capacities for compassionate relationships. *Companionship attachment* offers an intellectual friendship, a cognitive sharing that fosters creative imagination.

Nurturing caregiving in early life fosters optimal brain development, including of the prefrontal cortex critical for moral imagination (Schore, 2003a, b). Imaginative capacities in adults involve tacit knowledge, a trust in process, and an indwelling in the other, whether object or person (Polanyi, 1958). Living through the mind of the "other" involves an extended self. Moral imaginative capacities emerge from social creativity, based on these intensive social experiences in early life (although there are other sensitive periods in life when the brain can be reshaped

Table 1.1 Basic mindsets in triune ethics theory

Basic mindsets	Deliberative elaboration
Socially self-protective	
Safety: Bunker morality (aggression)	Vicious imagination
Safety: Wallflower morality (withdrawal, appeasement)	Detached imagination
Socially open	
Engagement ethic (relational attunement)	Communal imagination

to some degree). Those who have responsive caregivers, whose needs are met without distress, are more likely to develop secure attachment and the neurobiological underpinnings of a socially adaptive personality and moral intelligence (Eisenberg, 2000; Narvaez & Gleason, 2013). This personality is represented by the capacity for an engagement ethic and relational attunement with compassionate capabilities (Narvaez, 2008, 2012, 2014). When deliberative capacities are added to this base of relational attunement capacities, Communal imagination can flourish. Communal imagination uses capacities for abstraction from the present moment, addressing moral concerns beyond the immediate but grounded in a relational web, based in well-honed social skills. This broad sense of community was displayed by our hunter-gatherer cousins who were concerned with the welfare of all life forms, even into future generations (see Narvaez, 2013a).

One of the defining characteristics of the moral imagination is its ability to abstract and move beyond the present situation. This allows one to act on behalf of those not present or on behalf of abstract ideas like justice (Narvaez, 2010). The capacity to act on behalf of these abstract ideas requires the coordination of reasoning with motivational and emotional processes. For Communal imagination, empathy is a powerful source of moral behavior. When empathy-arousing stimuli are not present, the powers of imagination can still maintain their engagement in moral behavior. It is believed that individuals who demonstrate long-term commitments to humanitarian or prosocial causes rely on an ethic of imagination by making moral concerns central to their identity, and by selecting or seeking out situations that arouse their motivation and empathy to take action (Heath & Heath, 2010; Pizarro, 2000). In contrast, failure to help others commonly occurs because empathy is not engaged (Trout, 2009). For individuals who do not imaginatively regulate and heighten their emotional responses adaptively, "sympathy is easily aroused but quickly forgotten" (Wilson, 1993).

Those with poor early care, however, are likely to develop stress-reactive brains, making social interaction and an engagement ethic difficult to attain. Stress reactivity leads to a habitual safety ethic, shifting between different inegalitarian social orientations: an aggressive stance (bunker morality) or a withdrawal stance (wallflower morality). In those with sufficient physiological but less than optimal social experience in early life, the right brain is often underdeveloped, leading to a dominance of left-brain use (Schore, 1994, 2003a, b; Siegel, 1999). In this case, imagination can be divorced from compassion, resulting in the calculation of utility in Detached imagination (emotional disengagement) or the adoption of a non-imaginative ideology reflected in Vicious imagination (inegalitarian relations). This process is illustrated in Figure 1.1.

Moral imagination can be handicapped by violations of evolved childrearing practices that match the maturational schedule of the infant and young child (Narvaez, 2014). The developmental niche that evolved for humans in early life includes frequent on-demand breastfeeding for two to five years; nearly constant touch in the first years of life; responsiveness to the needs of the child so that the child does not become distressed; free self-directed play; multiple adult caregivers; positive expectation and support; and natural childbirth (Narvaez, Panksepp et al., 2013). Such evolved caregiving practices have been culturally discouraged, perhaps due to ignorance about their deep influence on development and lifelong capacities. "Undercare"—when these components are missing—undermines not only cognitive and emotional intelligence, but also moral creativity, for it is often the social-emotional systems that are underdeveloped as a result of modern caregiving practices (Schore 1994, 2003a, b; Trevarthen, 2005). Caregiving practices that violate evolved, expected care harm the capacities for moral engagement and Communal imagination and encourage the use of Detached and Vicious imagination, both self-focused uses of imagination (Narvaez, 2012, 2013, 2014).

Moral reasoning can be misused in two ways. First, when moral reasoning is calculative and divorced from relational empathy, imagination is limited, as reasoning seeks to apply a rule to a situation (Detached imagination). Calculative moral reasoning is harmful to moral imagination and action because it detaches from lived emotional experience and disengages social emotions. Actions originally viewed as immoral or even unthinkable can be justified among individuals who are detached from their prosocial emotions or are not experiencing empathy (Bandura, 1999). The road to habitual Detached imagination may be lubricated with poor social intuition or emotional intelligence.

A second form of reasoning misuse occurs when individuals or groups develop Vicious imagination. Vicious imagination seeks overt or covert dominance and control over others, demonstrating the superiority of the individual or group (for example, in terms of lifestyle, ideas, values, efficiency). It also is detached from empathy, but can be fueled by anger. In this case, moral reasons are used to justify actions or to confirm bias and strengthen preformed conclusions about inegalitarian relations. In extreme cases, individuals view human lives as secondary to their ends, and take evil action in a misguided effort to do good

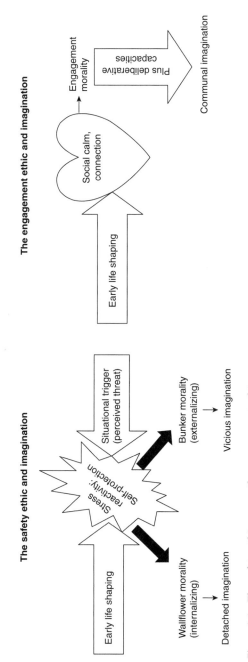

Figure 1.1 The safety ethic versus the engagement ethic.

(Bandura, 1999). As examples of pathological altruism, Baumeister and Vohs (2004) cite the Stalinist purges in Russia and the Cultural Revolution in China. In these and similar cases, the desire to "do good" was responsible for more deaths than actions considered "necessary evils" or based on revenge.

Everyday moral imagination

Early experiences, and reciprocal interactions with caregivers in particular, have a tremendous influence on the moral orientations that individuals develop. These moral orientations, in turn, influence everyday moral functioning. Multiple capacities are needed to respond to everyday moral situations with flexibility and imagination (Bartels, 2008). This section will address how individuals select goals and actions, develop habits, integrate numerous values into a single decision, and make sense of their actions and their identity in retrospect.

Moral focus

Imagination guides us in the selection of goals and action. For deep moral imagination:

> We need to imagine how various actions open to us might alter our self-identity, modify our commitments, change our relationships, and affect the lives of others. We need to explore imaginatively what it might mean, in terms of possibilities for enhanced meaning and relationships, for us to perform this or that action. We need the ability to imagine and to enact transformations in our moral understanding, our character, and our behavior. In short, we need an imaginative rationality that is at once insightful, critical, exploratory, and transformative. (Johnson, 1993, p. 187)

Imagining how an action might turn out facilitates choices and the eventual taking of action. The achievement of goals, such as sending letters, dieting, exercising, and even performance on a helping behavior, increases among those who imagine the actions they must take to achieve their desired outcome (Gollwitzer & Sheeran, 2006).

Emotional experiences, imaginative or real, can alter our judgments. Many of the abstract moral principles we come to endorse were formed as a result of an emotional experience that altered our judgment. In some situations, we feel empathy for individuals whom we judge negatively, and we change our higher-order principles as a result of our emotions in what Pizarro (2000) calls "bottom-up correction." For example, Batson et al. (1997) found that individuals had negative attitudes toward stigmatized groups such as the homeless, but revised those attitudes after feeling empathy for the group. Learning about a homeless man's experiences that led to his present condition resulted in greater sympathy the next time he was seen on the street (see Betancourt, Hardin, & Manzi, 1992). Imagining and understanding another's reality can change how we think and may even instigate an investigation so that we can understand the cause more deeply. An imagining individual uses abstraction capabilities with emotions engaged, becoming open to changing his or her thinking as a result of a dramatic mental experience. The many pitfalls of making judgments can be minimized with an integration of emotion and reasoned judgment, resulting in helpful assistance directed to those who will be benefited most (Loewenstein & Small, 2007). Such integration can occur among those with greater moral expertise (Narvaez, 2006, 2010; Narvaez & Lapsley, 2005).

Thinking in only abstract, philosophical terms leads to inferior moral decisions. Just as a person wanting to learn to drive a stick shift would not accomplish anything by pondering the matter outside of the car, we cannot learn much about moral action through detached or dispassionate thought. We must practice manipulating the stick shift and clutch within the process of driving. Similarly, moral imagination and action take place in the stream of life. As we cooperate with others, we learn how to perform positive moral actions. In Dewey's view (2009[1908]), it is vital to think interactively and examine moral behavior in light of its effects on relationships.

Habits

Habits are formed from immersion in environments that provide feedback on what works to meet one's aims or needs (Hogarth, 2001). Immersion trains implicit knowledge and automatic responses. It is best to choose environments that shape the intuitions and habits one wishes to have (Narvaez, 2006). During moral action, and reflection afterward, one gains a wealth of experiential or implicit knowledge to use in similar moral situations in the future (Narvaez & Lapsley, 2005).

Creative integration

Creative individuals draw on the experiences and successes of others. Exceptionally moral individuals are able to see the "bright spots" of what is currently helping people, and realize how they can use this knowledge to help in new and larger ways (Heath & Heath, 2010). For example, members of a non-governmental organization (NGO) with a small budget found that families in one community had the same low amount of money as those in other communities, but that this group did not suffer from stunted growth or malnutrition because they took advantage of key foods and cooking techniques. The NGO noticed this "bright spot," imagined its application elsewhere, and educated other communities about the same techniques. Drawing on prior successes cooperating with people with different and conflicting values, imaginative individuals are capable of using a number of values as they reason about issues (Tetlock, 2005), reconcile multiple considerations (Wallace, 1988), and take into account their responsibilities (Frankfurt, 1993).

Both TET and Dewey's theory of moral imagination address the importance of community, individual autonomy, and finding harmony between competing values (Fesmire, 2003; Narvaez, 2008). During the highest forms of moral imagination there is a double aim of valuing community-wide interests and maintaining respect for individual autonomy, rather than pitting one against the other (Rest et al., 1999). Frimer et al. (2011) found that moral exemplars who were especially altruistic and influential were able to act in accordance with values of agency and communion in the same actions, rather than favoring one or the other. Individuals who develop moral complexity and imagination are able to see a greater number of values as relevant to a situation rather than letting one override the others (see Baron & Spranca, 1997). They reason with complexity and see opportunities to fulfill multiple values at once, perceiving ways for values to be harmonious rather than in conflict (Narvaez, 2010).

Reflection

Reflection abilities develop from guided practice within particular domains. Taking time to consider routine behavior or analyze chosen actions facilitates further understanding. Through continued reflection, the growth process continues well after an action is completed. As implicit knowledge develops, action can become more automatic. People make attributions of responsibility and blame, evaluate the quality of decisions (Blum, 1994), and make sense of self-identity in light of behavior (Wainryb & Pasupathi, 2010), which alters perspectives the next time a similar situation occurs. Throughout the reflection process, imagination allows individuals to see opportunities to shape self-identity through action, to act in accordance with deeply held values, and to establish a more developed self as a result. TET posits the ability to frame behavior and establish a life narrative based on one's goals as an aspect of the moral imagination (Narvaez, 2010).

Creativity, intelligence, and moral imagination

Individual goal preferences are influenced by early experience: how open to others, how self-protective, how capable of thinking and reflecting the person is, and so on. Even beyond childhood, individuals are influenced by their culture and social context, and many seek and reflect on the aspects of culture and the environments that influence them (Pizarro, Detweiler-Bedell, & Bloom, 2006). However, other individuals do not reflect, but stay with learned habits and traditions within a small sphere, relying instead on others to tell them how to think or behave. In a morally pluralistic society, some individuals select moral goals, principles, and virtues that they wish to enact from a large variety of possibilities. In turn, individuals modify their views through interactions with others, whether parents or acquaintances and whether the idea is mainstream or radical. Pizarro, Detweiler-Bedell, and Bloom (2006) describe how individuals often do not passively accept the moral views of culture or parents, and how even a book or an interaction with a stranger can lead to a dramatic change in moral beliefs, especially among children who are reflective or imaginative.

The opportunity to step outside of the usual boxes of habit or intellectual detachment can engage the imagination, opening attention so that one can look at the world with fresh eyes (Hadot, 2011). For Communal imagination, this opportunity means adopting a "heart" view, engaging a sense of emotional connection to others (right-brain dominance), rather than using the filter of "utility" to narrow it (left-brain dominance). Reliance on rigid formulas, inflexible rules, or impersonal reasoning is, in Dewey's view and ours, "the death of all high moral responsibility" (Dewey, 2009[1980], 60). Instead, moral imagination requires an avoidance of simplistic thinking and a degree of ideological complexity. In research on both creativity and political ideologies, thinking that includes a strong need for closure, ambiguity intolerance, and dogmatism leads to less adequate decision-making. For example, need for closure has been linked with both lower creativity and more conformist, authoritarian moral ideologies. Ambiguity tolerance is consistently correlated with creativity and is also important in generating a morality that is not overly simplistic, reductive, and idealistic in viewing values as never conflicting (Tegano, 1990; Yurtsever, 2000).

In several studies, we have examined the explicit adoption of characteristics representing safety, engagement, or Communal imagination ethics. Unlike the few findings linking creativity and moral behavior, Communal imagination relates positively to a variety of moral characteristics and behaviors. These include honesty, integrity, empathy, perspective-taking, prosocial moral identity, action for the less fortunate, humanism, openness to experience, and a growth mindset (Narvaez, Brooks, & Hardy, 2013; Narvaez, Brooks, & Mattan, 2011). It should be emphasized that the relationship between Communal imagination and both honesty and integrity reveals a different picture than that portrayed by Ariely (2012; Gino & Ariely, 2012). Even if some measures of creativity are linked to cheating and poor integrity, Communal imagination is clearly not. It is also notable that Communal imagination is linked not only to judgment and personality measures (integrity) and thinking measures (humanism), but also to behavior measures (action for the less fortunate) and emotional measures (empathy). Further investigations are needed to determine whether these relationships are causal.

Creative moral exemplars

It is clear that moral creativity does not matter much unless an individual is able to choose one of the most useful ideas from those generated and act on it. Capacities for moral sensitivity, moral motivation, and follow-through must also be cultivated (Narvaez & Rest, 1995). Rest (1983, 1986) perhaps best captured the complete picture of moral functioning in his four-component model. He argued that moral sensitivity, moral judgment, moral motivation, and moral action each play important roles in moral decisions. Moral reasoning alone does not capture the whole picture. Moral sensitivity entails moral perception and interpretation, the ability to notice and identify the important ethical aspects of a situation. Moral judgment involves choosing the morally ideal course through reasoning. Moral motivation prioritizes one moral action over other options. Moral action involves the ability and character to act, through will and knowledge, on one's moral goals and judgments.

The creative moral exemplar possesses a variety of skills that function as a toolkit. The skills mentioned throughout this chapter fit into one or more of Rest's components. Emotion to feel with others and perceive their needs fits best into moral sensitivity. The abilities to select goals and values to endorse, integrate numerous values into a single decision, and generate several ideas about how to act fit best into moral judgment. The use of emotion and regulation techniques to maintain motivation, focus on a problem, and perhaps form one's identity in light of actions fits into moral motivation. Finally, the skills of moral action include cultivating good habits and selecting among the opportunities or values one can apply.

Individuals high in moral imagination are more likely to extend regard to members of outgroups or strangers (Mrkva & Narvaez, in press). They are less likely to stigmatize the homeless based on information suggesting responsibility for their condition, more likely to favor policies that promote greater respect for each life (whether in the USA or abroad, for example), and less likely to blame victims or ignore individuals in need because of their status as strangers or an "other."

Many of these components are included in the right-brain capacity for mindfulness, a flexible engagement in the present, the ability to see connections, sensitivity to context, and noticing new elements of a situation. Mindfulness requires creativity as well as engagement in the present moment, sensitivity to others in their immediate environment, and willingness to interact with and help others, if the feelings and actions of others suggest that they are in need or could be assisted in some way. Mindfulness entails the ability to empathize with others and experience their emotions, but imaginative moral functioning must also use this experience to guide changes in thought. In this way, mindfulness can influence moral reasoning, judgment, and action as much as sensitivity, in a bottom-up correction (Pizarro, 2000). We can see this in a trend identified by Ray and Anderson (2000): the emergence of "cultural creatives," individuals who demonstrate characteristics that blend morality and creative imagination (see Table 1.2).

Conclusion

Imagination as a mental faculty, emotion as a psycho-bio-social faculty, and morality as an internalized mental frame interact developmentally in different ways based on a person's experiences. These differential interactions can result in different dispositions and types of imagination: Detached (little emotional engagement with the world), Vicious (aggressive emotional interaction), Engaged (present-focused positive interaction), and Communal (extended collaborative, positive interaction). Creativity is demonstrated to different degrees and at varying levels of quality in each of these types of moral imagination. Although high levels of both intelligence and creativity may be demonstrated in

Table 1.2   Characteristics of cultural creatives

Care deeply about the natural world
Awareness of and desire for action on planet-wide issues (global warming, poverty, and so on)
Activism for positive social change
Willingness to pay higher taxes for the benefit of the environment
Value developing and maintaining relationships
Value helping others
Volunteer for good cause
Value spirituality (but fear fundamentalism)
Value spiritual and psychological development
Value equality for women in all spheres
Concerned for the well-being of women and children
Want government to focus on education, welfare, and sustainable living
Unhappy with left and right politics
Optimistic about the future
Desire to create better way of life for all
Concerned about corporate profit motives and destructive side effects
Unlikely to overspend or be in debt
Deplore emphasis on consumption, status, and monetary success
Enjoy exotic people and places

Source: Ray & Anderson, 2000.

Detached and Vicious imagination, these forms are more self-focused, limited in their scope of care and consequences for others, and lead to intentional or unintentional harmful outcomes. Communal imagination, in contrast, maintains an inclusive sense of caring in pondering and taking action as a creative collaboration (John-Steiner, 2000), demonstrating the highest form of ethical sensitivity. However, we must not claim that imagination is used in all moral decisions. There are many occasions when we make moral judgments based on habit or expediency, and fail to consider the uniqueness of the situation. If we are correct about the significant role creativity plays in our moral lives, imagination—especially social imagination—deserves more attention from psychological researchers and ethical theorists.

Morality is "the ongoing imaginative exploration of possibilities for dealing with our problems, enhancing the quality of our communal relations, and forming significant personal attachments that grow" (Johnson, 1993, p. 209). Current cultural practices do not well serve the development of the more positive forms of moral imagination. There are more supports for Detached and Vicious imagination than for Engaged and Communal imagination, including forms that are societal (for example, undercare, priority of monetary success) and educational (schooling that sets aside the emotional and social aspects of life). Engaged and Communal moral imagination require good beginnings, with nurturing caregiving and empathic relationships during sensitive periods. These experiences foster right-brain, present-oriented capacities (including self-regulation, behavior inhibition, and empathy; Narvaez, Wang et al., 2013). Engaged and Communal imagination also may require ongoing safe and supportive environments. Creative moral imagination allows individuals and communities to grow in virtue, deepening and extending moral regard and sensitivity to a greater circle of life.

References

Ariely, D. (2012). *The (Honest) Truth about Dishonesty: How We Lie to Everyone— Especially Ourselves*. New York: HarperCollins.

Bandura, A. (1999). Moral disengagement in the perpetration of inhumanities. *Personality and Social Psychology Review, 3*, 193–209.

Baron, J., & Spranca, M. (1997). Protected values. *Organizational Behavior and Human Decision Processes, 70*, 1–16.

Bartels, D. M. (2008). Principled moral sentiment and the flexibility of moral judgment and decision making. *Cognition*, 108(2), 381–417.

Batson, C. D., Polycarpou, M. P., Harmon-Jones, E., Imhoff, H., Mitchener, E. C., Bednar, L. L., Klein, T. R., & Highberger, L. (1997). Empathy and attitudes: Can feeling for a member of a stigmatized group improve feelings toward the group? *Journal of Personality and Social Psychology, 72*, 105–118.

Baumeister, R. F., & Vohs, K. D. (2004). Four roots of evil. In A. G. Miller (ed.), *The Social Psychology of Good and Evil* (pp. 85–101). New York: Guilford.

Ben-Ze'ev, A. (2000). *The Subtlety of Emotions*. Cambridge, MA: MIT Press.

Betancourt, H., Hardin, C., & Manzi, I. J. (1992). Beliefs, value orientation, and culture in attribution processes and helping behavior. *Journal of Cross-Cultural Psychology, 23*, 179–195.

Blasi, A. (1999). Emotions and moral motivation. *Journal for the Theory of Social Behavior, 29*, 1–19.

Blum, L. (1994). *Moral Perception and Particularity*. New York: Cambridge University Press.

Dewey, J. (2009[1908]). *Ethics*. Ithaca, NY: Cornell University Press.

Eisenberg, N. (2000). Emotion, regulation, and moral development. *Annual Review of Psychology, 51*, 665–697.

Fesmire, S. (2003). *John Dewey and the Moral Imagination: Pragmatism in Ethics*. Bloomington: Indiana University Press.

Frankfurt, H. (1993). What we are morally responsible for. In J. M. Fischer & M. Ravizza (eds.), *Perspectives on Moral Responsibility* (pp. 286–294). Ithaca, NY: Cornell University Press.

Fredrickson, B. L., & Branigan, C. (2005). Positive emotions broaden the scope of attention and thought-action repertoires. *Cognition & Emotion, 19*(3), 313–332.

Frimer, J. A., Walker, L. J., Dunlop, W. L., Lee, B. H., & Riches, A. (2011). The integration of agency and communion in moral personality: Evidence of enlightened self-interest. *Journal of Personality and Social Psychology, 101,* 149–163.

Gibbs, J. C. (2003). *Moral Development and Reality: Beyond the Theories of Kohlberg and Hoffman.* Thousand Oaks, CA: Sage.

Gino, F., & Ariely, D. (2012). The dark side of creativity: Original thinkers can be more dishonest. *Journal of Personality and Social Psychology, 102,* 445–459.

Gollwitzer, P. M., & Sheeran, P. (2006). Implementation intentions and goal achievement: A meta-analysis of effects and processes. *Advances in Experimental Social Psychology, 38,* 69–119.

Greenspan, S. I., & Shanker, S. I. (2004). *The First Idea.* Cambridge, MA: Da Capo Press.

Hadot, P. (2011). *The Present Alone Is Our Happiness* (M. Djaballah & M. Chase, trans.). Stanford, CA: Stanford University Press.

Haidt, J. (2001). The emotional dog and its rational tail: A social intuitionist approach to moral judgment. *Psychological Review, 108,* 814–834.

Haidt, J. (2007). The new synthesis in moral psychology. *Science, 316,* 998–1002.

Heath, C., & Heath, D. (2010). *Switch: How to Change Things When Change Is Hard.* New York: Broadway Books.

Hogarth, R. M. (2001). *Educating Intuition.* Chicago, IL: University of Chicago Press.

Hume, D. (1969[1739/1740]). *A Treatise of Human Nature.* London: Penguin.

Isen, A. M., & Daubman, K. A. (1984). The influence of affect on categorization. *Journal of Personality and Social Psychology, 47*(6), 1206.

John-Steiner, V. (2000). *Creative Collaboration.* New York: Oxford University Press.

Johnson, M. (1993). *Moral Imagination.* Chicago, IL: University of Chicago Press.

Kant, I. (1949). *Fundamental Principles of the Metaphysics of Morals.* New York: Liberal Arts Press.

Klinger, E. (1978). Modes of normal conscious flow. In K. S. Pope & J. L. Singer (eds.), *The Stream of Consciousness: Scientific Investigations into the Flow of Human Experience* (pp. 225–258). New York: Plenum.

Loewenstein, G., & Small, D. (2007). The scarecrow and the tin man: The vicissitudes of human sympathy and caring. *Review of General Psychology, 11,* 112–126.

Lupien, S. J., McEwen, B. S., Gunnar, M. R., & Heim, C. (2009). Effects of stress throughout the lifespan on the brain, behaviour and cognition. *Nature Reviews Neurosciences, 10,* 434–445.

MacLean, P. D. (1990). *The Triune Brain in Evolution: Role in Paleocerebral Functions.* New York: Plenum.

McGilchrist, I. (2009). *The Master and His Emissary: The Divided Brain and the Making of the Western World.* New Haven, CT: Yale University Press.

Meaney, M. (2010). Epigenetics and the biological definition of gene × environment interactions. *Child Development, 81*(1), 41–79.

Monin, B., Pizarro, D., & Beer, J. (2007). Emotion and reason in moral judgment: Different prototypes lead to different theories. In K. D. Vohs, R. F. Baumeister, & G. Loewenstein (eds.), *Do Emotions Help or Hurt Decision Making?* New York: Russell Sage Foundation Press.

Mrkva, K., & Narvaez, D. (in press). Moral psychology and the "cultural other": Cultivating an openness to experience and the new. In B. Zizek & A. Escher (eds.), *Ways of Approaching the Strange*. Stuttgart: Franz Steiner Verlag.

Murdoch, I. (1989[1970]). *The Sovereignty of Good*. London: Routledge.

Narvaez, D. (2006). Integrative ethical education. In M. Killen & J. Smetana (eds.), *Handbook of Moral Development* (pp. 703–733). Mahwah, NJ: Lawrence Erlbaum Associates.

Narvaez, D. (2008). Triune ethics: The neurobiological roots of our multiple moralities. *New Ideas in Psychology, 26*, 95–119.

Narvaez, D. (2010). Moral complexity: The fatal attraction of truthiness and the importance of mature moral functioning. *Perspectives on Psychological Science, 5*, 163–181.

Narvaez, D. (2012). Moral neuroeducation from early life through the lifespan. *Neuroethics, 5*(2), 145–157. doi:10.1007/s12152-011-9117-5.

Narvaez, D. (2013a). Development and socialization within an evolutionary context: Growing up to become "A good and useful human being." In D. Fry (ed.), *War, Peace and Human Nature: The Convergence of Evolutionary and Cultural Views* (pp. 643–672). New York: Oxford University Press.

Narvaez, D. (2013b). Neurobiology and moral mindsets. In K. Heinrichs & F. Oser (eds.), *Moral and Immoral Behavior: Theoretical and Empirical Perspectives on Moral Motivation* (pp. 289–307). Rotterdam: Sense Publishers.

Narvaez, D. (2014). *The Neurobiology and Development of Human Morality: Evolution, Culture and Wisdom*. New York: W.W. Norton.

Narvaez, D., & Gleason, T. (2013). Developmental optimization. In D. Narvaez, J. Panksepp, A. Schore, & T. Gleason (eds.), *Evolution, Early Experience and Human Development: From Research to Practice and Policy* (pp. 307–325). New York: Oxford University Press.

Narvaez, D. & Lapsley, D. K. (2005). The psychological foundations of everyday morality and moral expertise. In D. Lapsley & C. Power (eds.), *Character Psychology and Character Education* (pp. 140–165). Notre Dame, IN: University of Notre Dame Press.

Narvaez, D., & Rest, J. (1995). The four components of acting morally. In W. Kurtines & J. Gewirtz (eds.), *Moral Behavior and Moral Development: An Introduction* (pp. 385–400). New York: McGraw-Hill.

Narvaez, D., & Vaydich, J. (2008). Moral development and behavior under the spotlight of the neurobiological sciences. *Journal of Moral Education, 37*, 289–313.

Narvaez, D., Brooks, J., & Mattan, B. (2011). Triune ethics moral identities are shaped by attachment, personality factors and influence moral behavior. Annual meeting of the Society for Personality and Social Psychology, San Antonio.

Narvaez, D., Hardy, S., & Brooks, J. (2014). A multidimensional approach to moral identity: Early life experience, prosocial personality, and moral outcomes. Manuscript submitted for publication.

Narvaez, D., Panksepp, J., Schore, A., & Gleason, T. (eds.) (2013). *Evolution, Early Experience and Human Development: From Research to Practice and Policy*. New York: Oxford University Press.

Narvaez, D., Wang, L., Gleason, T., Cheng, A., Lefever, J., & Deng, L. (2013). The evolved developmental niche and sociomoral outcomes in Chinese three-year-olds. *European Journal of Developmental Psychology, 10*(2), 106–127.

Panksepp, J. (1998). *Affective Neuroscience*. New York: Oxford University Press.

Pizarro, D. (2000). Nothing more than feelings? The role of emotions in moral judgment. *Journal for the Theory of Social Behaviour, 30*, 355–455.

Pizarro, D., Detweiler-Bedell, B., & Bloom, P. (2006). The creativity of everyday reasoning: Empathy, disgust and moral persuasion. In J. Kaufman & J. Baer (eds.), *Creativity and Reason in Cognitive Development*, (pp. 81–98). Cambridge: Cambridge University Press.

Polanyi, M. (1958). *Personal Knowledge: Towards a Post-Critical Philosophy*. Chicago, IL: University of Chicago Press.

Porges, S. W. (2011). *The Polyvagal Theory: Neurophysiologial Foundations of Emotions, Attachment, Communication, Self-Regulation*. New York: Norton.

Ray, P. H., & Anderson, S. R. (2000). *The Cultural Creatives: How 50 Million People Are Changing the World*. New York: Harmony Books.

Reid, S. (1999). Autism and trauma. Autistic post-traumatic developmental disorder. In A. Alvarez & S. Reid (eds.), *Autism and Personality. Findings from the Tavistock Autism Workshop* (pp. 93–109). London: Routledge.

Rest, J. (1983). Morality. In P. H. Mussen (series ed.) & J. Flavell & E. Markman (vol. eds.), *Handbook of Child Psychology: Vol. 3, Cognitive Development* (4th edn., pp. 556–629). New York: John Wiley & Sons.

Rest, J. R. (1986). *Moral Development: Advances in Research and Theory*. New York: Praeger.

Rest, J., Narvaez, D., Bebeau, M. J., & Thoma, S. J. (1999). *Postconventional Moral Thinking: A Neo-Kohlbergian Approach*. Mahwah, NJ: Lawrence Erlbaum Associates.

Sapolsky, R. (2004). *Why Zebras Don't Get Ulcers*, 3rd edn. New York: Holt.

Schore, A. N. (1994). *Affect Regulation and the Origin of the Self*. Hillsdale, NJ: Lawrence Erlbaum Associates.

Schore, A. (2003a). *Affect Regulation and the Repair of the Self*. New York: Norton.

Schore, A. (2003b). *Affect Dysregulation and Disorders of the Self*. New York: Norton.

Shelley, P. B. (1821). *A Defence of Poetry*. Downloaded on September 16, 2012 from http://www.bartleby.com/27/23.html.

Shonkoff, J. P., & Garner, A. S., The Committee on Psychosocial Childhood, Adoption, and Dependent Care, and Section on Developmental and Behavioral Pediatrics, Siegel, B. S., Dobbins, M. I., Earls, M. F., Garner, A. S., McGuinn, L., Pascoe, D., & Wood, D. L. (2012). The lifelong effects of early childhood adversity and toxic stress. *Pediatrics, 129*, e232. doi: 10.1542/peds.2011-2663.

Siegel, D. J. (1999). *The Developing Mind: How Relationships and the Brain Interact to Shape Who We Are*. New York: Guilford.

Slade, A. (1987). The quality of attachment and symbolic play. *Developmental Psychology, 23*, 78–85.

Slade, A. (1994). Making meaning and making believe: Their role in the clinical process. In A. Slade & D. Wolf (eds.), *Children at Play: Clinical and Developmental Approaches to Meaning and Representation* (pp. 81–110). New York: Oxford University Press.

Slovic, P., Finucane, M., Peters, E., & MacGregor, D. G. (2002). Rational actors or rational fools: Implications of the affect heuristic for behavioral economics. *Journal of Socioeconomics, 31*(4), 329–342.

Somerville, M. (2006). *The Ethical Imagination: Journeys of the Human Spirit.* Toronto: House of Anansi Press.

Sternberg, R. J. (ed.). (1999). *Handbook of Creativity.* New York: Cambridge University Press.

Tegano, D. W. (1990). Relationship of tolerance of ambiguity and playfulness to creativity. *Psychological Reports, 66*(3), 1047–1056.

Tetlock, P. E. (2005). *Expert Political Judgment: How Good Is It? How Can We Know?* Princeton, NJ: Princeton University Press.

Trevarthen, C. (2005). Action and emotion in development of the human self, its sociability and cultural intelligence: Why infants have feelings like ours. In J. Nadel & D. Muir (eds.), *Emotional Development* (pp. 61–91). Oxford: Oxford University Press.

Trevathan, W. R. (2011). *Human Birth: An Evolutionary Perspective*, 2nd edn. New York: Aldine de Gruyter.

Trout, J. D. (2009). *The Empathy Gap.* New York: Viking/Penguin.

Wainryb, C., & Pasupathi, M. (2010). Political violence and disruptions in the development of moral agency. *Child Development Perspectives, 4*, 48–54.

Wallace, J. D. (1988). *Moral Relevance and Moral Conflict.* Ithaca, NY: Cornell University Press.

Wilson, J. Q. (1993). *The Moral Sense.* New York: Free Press.

Yurtsever, G. (2000). Ethical beliefs and tolerance of ambiguity. *Social Behavior and Personality, 28*(2), 141–148.

2
Moral Craftsmanship

Mark Coeckelbergh
University of Twente, The Netherlands

Not all art is morally acceptable. For example, most people would agree that it is wrong to torture a person for the sake of an *art performance* or to kill a person for *aesthetic* reasons. More generally, if something qualifies as "beautiful," that does not necessarily make it *right* or *good*. At the same time, the boundary between ethics and aesthetics may not be as clear as many people think. Can morality be strictly separated from aesthetics? Perhaps ethics itself already has an "aesthetic" dimension. One of the questions in this volume concerns how people can be creative *in* morality and how they can change the ethics of their lives, practices, and societies. Can morality itself be creative? What does it mean to be "morally creative"?

In this chapter, I briefly explore what moral creativity could mean by developing the notion of "moral craftsmanship." What does "craftsmanship" mean in the moral domain? What kind of creativity is it? What kind of learning does it involve? What might be an example of moral craftsmanship? And what can be gained by using this notion?

I start with the notion of craftsmanship and analyze the kind of knowledge, creativity, imagination, and learning it requires. In particular, I show that craftsmanship involves the development of know-how and skills, and of openness to and engagement with the world and with others. Then I argue that the moral creativity we need to cope with problems in the real world requires a similar kind of knowledge, creativity, imagination, and learning. I show how this view calls into question Platonic and modern conceptions of moral knowledge, and how it suggests a view of moral change that emphasizes personal and societal growth rather than "moral design."

Craftsmanship and learning skills

A craftsperson does not only "do a job" and "get the work done," but also works on something with great skill, which he or she has attained through training. Furthermore, craftsmanship typically involves physical and bodily engagement with things and tools. This gives the craftsperson a particular kind of knowledge. In *The Craftsman*, Sennett (2008) gives examples of craftsmanship ranging from cooking to medicine and music. He argues that these physical, bodily practices give us tactile experience and relational understanding: a "tacit knowledge" (2008, p. 50). Someone who is a craftsperson knows how, rather than only knows that. Those who have this kind of knowledge do good work and achieve excellence in their work. Achieving excellence is also something that matters to the craftsperson: he or she is committed and motivated to creating something excellent.

The kind of creativity, imagination, and learning involved in this work is not so much about thinking up a concept, which is then applied to the matter at hand, but also and mainly about creating by handling, about improvising, and about learning while doing. For example, a creative cook may read a recipe, but during cooking he or she uses know-how and experience, acquired during years of learning and practice, to create a meal that is so much more than the recipe and that improvises on the recipe in original and unpredictable ways. This kind of cook does not first imagine the meal and then start working to create or recreate that mental image. Excellence in craftsmanship is not the result of a new "concept," but rather emerges from and in the process of handling. It is a tacit kind of knowledge, since it cannot completely be turned into a recipe. It has developed by means of tactile experience and involvement with food and people. If it is imagination at all, then it is not imagination as the production of a mental image, but rather the very practical creation and improvisation that go on during the crafting.

This does not mean that explicit instruction is obsolete. In the beginning, the apprentice requires explicit instruction. The apprentice cook needs a recipe and instructions for how to use things in the kitchen; the apprentice musician needs a score and instruction in how to play the instrument; and the apprentice medical surgeon needs instruction and theoretical knowledge about the human body.

This is true for all skilled learning. Consider how we learn to ride a bike, how we learn to drive a car, how we learn to fly an airplane. In order to become skilled in anything, we have to start with explicit instruction and some theory. But, as Dreyfus and Dreyfus (1991) have argued, in order to become experts, we need to acquire the tacit knowledge, the know-how, the knowledge-in-practice to become crafty at it. Moreover, acquiring this knowledge is impossible if we take a detached position and attitude. Instead, we need to handle the material and take up the tools (for example, get on the bike), immerse ourselves in the activity, persist in it and continue to train (also when, for example, we fall off the bike), and have the commitment to try to get better at it.

The craftsperson, then, is a skilled expert of the "highest" kind: he or she is not only competent, but also creative. The craftsperson has a kind of knowledge that enables him or her to cook, make music, perform surgery, and so on, in a way that, in a sense, no one else does and can do. The knowledge of the craftsperson has a "personal" dimension, since his or her particular creativity is linked to his or her particular way of engaging with things and people. There are usually many more craftspeople who are in the same craft or trade (for example, making a particular type of music, doing a specific category of surgery)—that is, *what* they all do is the same—but exactly what each of them does can never be reduced to this type or category. They are seen as reinventing it, each in their own way, as going beyond it, as creating their own category, since *how* they do it is unique. Their know-how is partly shared with others and partly unique. What they do is more imaginative or creative than others, but not because they are in possession of a concept, map, recipe, or code that potentially could easily be shared but that they keep to themselves for some reason. Their creativity comes from having acquired a particular kind of know-how. Such know-how can only partially be transmitted and can only be shared by means of a personal learning relation, by means of apprenticeship. However, the goal of the apprentice is not to copy the master; imitation is part of the learning, but only in the beginning. The goal is to attain his or her own *personal* craftsmanship that grows by means of a specific, personal training trajectory and within a learning relation with the master craftsperson.

Indeed, not only the learning but the whole practice of craftsmanship is a deeply social activity, as it was in medieval times and also today. Consider people working together in a car repair workshop: the craftspeople collaborate and there are relations of companionship between them. The learning that takes place is also social learning: they share knowledge and each develops expertise in relation to the others. Much of the know-how is shared by the whole, by the workshop. Again, this means that the creativity of craftsmanship involves a unique kind of knowledge that could only develop in *this* workshop in collaboration with *these* people. The garage workshop, the intensive care unit in a hospital, the kitchen of a restaurant, and so on are all places of craftsmanship, where people are gathered around tools and with other people to learn, to become more skilled, and to develop a creativity unique to the place and the team in which they work. The result of the expert craftsperson's creative work, then, can be called excellent in at least the following senses.

First, the product of the work is excellent, since it is the result of the highest creativity and expertise, based on long training in particular social and physical settings, related to particular people and tools. Second, the craftsperson also achieves personal excellence: he or she has become a good craftsperson. In both senses, the excellence achieved by the craftsperson rests on the "how," on the way toward excellence, the route that makes something excellent.

Moral craftsmanship

If this is an adequate description of craftsmanship, in particular the *creativity* and learning of craftsmanship, what could we learn from it for morality? What would craftsmanship mean in the moral domain? What is moral know-how, and what role does—and should—it play in morality? In what sense does moral learning consist in the learning of skills, personal learning, and social learning? What kind of excellence is produced?

In order to articulate the notion of moral craftsmanship, let me make a distinction between two conceptions of moral creativity, which are connected to two conceptions of moral imagination. One is Platonic and modern; the other is influenced by pragmatism, phenomenology, and Aristotelian thinking.

Moral creativity as conceptual design

A typical response to ethical problems—problems we have now or problems we expect in the future—is to call for rules and principles, a new moral code that should help to guide and regulate ethical conduct. This approach is often used by moral philosophers, many of whom are fond of moral principles and theories, but it is also typical of the way society and politics respond to problems. If there are problems in global finance, we call for ethical codes and more regulation. If there are problems in prostitution, we want new laws. And if we think about ethical problems with future autonomous robots, we try to make ethical laws for those robots: for instance, the laws proposed in Asimov's (1942) science fiction stories and in contemporary robot ethics that reflect on how to make them into "moral machines" (Wallach & Allen, 2009).

If we adopt this approach to morality, we assume the modern view that *humans* have to be made into moral machines whose moral subjectivity, if they have any at all, is limited to making or following laws. Society can only become more ethical if people follow moral rules. The rule-giver or law-giver is a perfectly rational agent who knows what is reasonable, and we have the choice either to become such a perfectly rational agent who is able to deliberate autonomously about what is right or—if we cannot—to follow the super-agent's rules.

The model that fits this moral epistemology is theoretical science (if a "pure" theoretical science exists at all), the practice of law-making, and perhaps also classical artificial intelligence. We philosophers, and perhaps all citizens who use their capacity of reasoning, conceive of rules in our study room and rationally think about which laws should govern society. For example, deontological philosophers propose moral and political rights and laws that should protect individuals. Utilitarian philosophers start from a concept and a calculus, and then try to (re)design society. Bentham (1970[1789]) proposed a calculus to determine which act would bring about the greatest happiness for the greatest number of individuals. In the twentieth century, Peter Singer (1975) proposed a utilitarian principle that is supposed to solve issues in animal ethics. At work is the philosopher as reasoner and as moral scientist who tries to figure out which moral principles or moral laws should apply.

Moral philosophers working in this tradition seek to change society from top down. In this respect, they do not differ from Plato, who envisioned the ideal of the philosopher-king designing the perfect, well-ordered society. Of course, their model is not science as we know it today. Plato did not take to experimental science as a method to gain knowledge. Instead, he held that gaining moral knowledge was a matter of *theoria*, which might be translated as "insight" or "vision." Yet, there is a sense in which both the Platonic and the modern way of understanding morality share the same approach to moral creativity: they think that such creativity is about *theoria*, about imagining and creating a blueprint of the human as moral (since perfectly rational) and as good, of the just society. The philosopher designs a concept of the right (modern ethics) or a concept of the good (ancient ethics). This requires a kind of imagination, but it is a detached mode of knowing, knowing and reasoning that go on "in my head." The philosopher contemplates the good, or reasons about the right, while looking at the world from a distance. It is what Nagel called a "view from nowhere" (Nagel, 1986).

Furthermore, this view presupposes that there is a fixed good, that once we have theory about the good, we only need to "apply" it to the messy, practical world. But is this kind of moral knowledge necessary and sufficient for morality? Is this kind of creativity and imagination sufficient for improving the moral quality of persons and of society? For example, is rational thinking enough to cope with technological risk, or do we also need emotions? And is "study-room" imagination enough to cope with the moral challenges we face as individuals and societies? For example, can principles alone help us to cope with complex and pressing global problems such as energy and water shortages, climate change and environmental degradation, financial and economic crisis, and military conflict? According to Popper, Plato's "wise men" are too occupied with the problems of the superior world; they hold fast to "the ordered and the measured" and have no time to "look down at the affairs of men" (Popper, 2013[1945], p. 138). Moral creativity, it seems, not only requires detached science and metaphysics, it also has something to do with emotions and intuitions, and there seems to be something problematic about thinking up moral rules and a moral order while residing in the proverbial ivory tower without engaging with practical problems.

Inspired by the Humean and pragmatist response to Kantian and utilitarian philosophy, one could try to combine principles and imagination by acknowledging that a different kind of imagination and emotions plays a role in moral reasoning. For example, one might think about the consequences of following a rule. One might project oneself into a particular situation. One could also use empathy, and ask what it would be like to be *that* particular person in *that* situation. Indeed, some positions in the literature on moral imagination (for example, Johnson, 1993; Coeckelbergh, 2007) accept that morality is basically about principles, but argue that it has an imaginative dimension. They try to combine the view that morality is basically a form of moral reasoning with the idea that imagination, emotions, intuition, and so on play an important role *in* that reasoning. This view is not wrong, but we also need to ask the question of whether morality is *more* than moral reasoning and *more* than thinking about principles. We need to move on to a more radical view of moral creativity, which rejects the basic assumptions of the Platonic and modern view. Such a view could run as follows: Morality is not all about reasoning, and not even all about imagination as a mental capacity or mental operation that acts as a kind of tool or plug-in for moral reasoning. Moral creativity and moral imagination must be understood in a more practical and social way, one that is part of a much more "fluid" view of morality. Let me now start to unpack and develop this intuition.

Perhaps we use our imagination when we consider the moral consequences of our actions or when we engage in moral reasoning by using emotions and empathy, as Humeans think we do. But in both cases, it is an imagination *within* reasoning: as moral reasoners and deliberators, we imagine the consequences of having particular rules, imagine how it would be to be that other person, and so on. This kind of creativity and imagination still goes on "in my head"; it has its origin in the Cartesian subject that is disconnected from the world. It is still about seeing the good from a distance. The view that morality is more a matter of empathy and of feeling, perception, intuition, and moral vision was and is a welcome response to the rationalistic tendencies in modern thinking. However, emphasizing emotions, empathy, and intuition remains a modern response if and insofar as it presupposes a non-relational moral subject, one who is not engaged with the world, contemplating morality in the Cartesian cocoon of his or her mind. In order to avoid concepts such as imagination or intuition being recuperated by the study-room model of moral thinking, they should be given a new role within a relational, non-Platonic, non-Cartesian, and non-Kantian view of morality and of knowledge. According to this view, acquiring moral knowledge is not only about achieving *know-that* (values, principles, rules), not about having a theoretical "intuition" or "vision" separated from the world, and not even only about imagining the consequences of one's actions and putting oneself in the other's shoes. Instead, it is about real-world *know-how*, feeling one's way through the world, about grasping, about handling, and about touching. For instance, when dealing with the current economic and financial crisis, we do not have the luxury to start from a blank slate. We are already in trouble, and we need to find creative solutions that start from the messy world in which we find ourselves and from the concrete problems we face—as individuals, but also as societies and as people who are called on to respond to others. The kind of moral creativity needed here is of a more *practical* kind.

In the next section, I will articulate a different conception of moral creativity and moral imagination by using instead models and metaphors from non-conceptual art, music, dance, cooking, gardening, and creative manual work. Inspired by Dewey's pragmatism (in particular Fesmire's interpretation of Dewey), Heideggerian phenomenology (for instance Dreyfus and Borgmann), Aristotelian thinking (MacIntyre), and accounts of craftsmanship (Sennett, Crawford), I will articulate the notion of "moral craftsmanship." This will imply a conception of morality that replaces the detached rational moral subject of modern philosophy by the involved subject, engaged in the world. The exercise of *that* kind of imagination neither relies on ivory-tower moral science nor Cartesian-style empathy, but rather emphasizes moral dancing, moral improvisation, moral engineering, and moral tinkering. Moreover, all these kinds of moral doings and moral know-how crucially involve others. We cannot fix moral problems alone and we have to stay in tune with others. I argue that the moral creativity and moral imagination we need are *both more practical and more social*. I suggest that a truly creative and imaginative ethics is not about sitting down and creating laws, mental images, theoretical visions, individual artistic "genius," or "concepts," which then have to be applied to the "real world" and provide guidelines for "ordinary" people. Rather, a truly creative and imaginative ethics involves the development and use of moral skill in order to cope with the moral problems in which all of us are involved, which requires a basic openness to, and engagement with, the world, as well as an openness to and engagement with *others*, on which our life and the solution to our moral problems depend.

Moral creativity as moral craftsmanship

A good starting place for articulating this alternative approach to moral creativity is Dewey's ethics. According to Dewey, ethics is not centrally about theory, but about solving practical problems. Let us begin with his moral epistemology in order to achieve some distance from the modern and Platonic approach.

In *Human Nature and Conduct*, Dewey distinguishes between knowing how and know that, and argues that we *"know how* by means of our habits"* (Dewey, 1922, p. 177; Dewey's emphasis). If we have the latter kind of knowledge, our knowledge is not theoretical or reasoned but a matter of practical skill and habit. The knowledge we need for ethics, then, is more about knowing how than knowing that: it is about knowing how to live a good life by developing practical skill and good habits. Habits are not "mental" as opposed to "non-mental," they involve skills, and skills are embodied, physical, and material. Dewey writes: "We should laugh at any one who said that he was master of stone working, but that the art was cooped up within himself and in no wise dependent upon support from objects and assistance from tools" (Dewey, 1922, p. 15).

Similarly, we should laugh at anyone who suggests that acquiring moral mastery is in no way bodily or material. As relational beings and as embodied moral subjects, we are already standing in relation to things and to others. Morality is not merely conceptual but arises out of "active connections of human beings with one another" (Dewey, 1922, p. 225); these connections are also related to our bodies and to matter. Let me further unpack this view of moral knowledge.

Heidegger also objected to modern Cartesian dualism. We are not reasoning egos separated from the world; we are always already engaged in the world. He used the term "being-in-the-world" (Heidegger, 1996[1927]) to denote this involved character of human existence. For moral knowledge, this view implies that we cannot achieve complete, sufficient moral knowledge as detached reasoners or as detached feelers. Both the rationalist view (for example, Kantian or utilitarian) and the Humean view presuppose that we are first detached, Cartesian moral subjects who then reason about moral problems and try to bridge the distance between ourselves and the world, and between ourselves and others. However, this modern idea of detached moral reasoning denies that we are beings-in-the-world, that our moral subjectivity is already relational from the start. Moreover, even if there were such a thing as detached reasoning, the phenomenology of moral experience shows that what we take to be explicit theoretical deliberation is an exceptional state. According to Dewey, the need for moral reflection only arises in terms of crisis, when habits no longer work (see also Pappas, 2008, p. 122). Normally morality is not a matter of detached reasoning but moral development-in-action. In particular, it is a matter of the development of skill. Let me say more about this developmental aspect of skill by using Dreyfus.

Dreyfus and Dreyfus (1991) argue that moral knowledge should be conceptualized in terms of skill. We are always already embodied and engaged in practices when we encounter a moral problem. Acquiring moral knowledge is a matter of building practical know-how. If we want to become morally mature, we need to learn from experience and use that know-how "so as to respond more appropriately to the demands of others in the concrete situation" (Dreyfus & Dreyfus, 1991, p. 247). If this requires intuition, it is intuition that is cured by experience and developed through experience. It is not a mental, theoretical exercise but a worldly, practical exercise. We could add that it better resembles how manual workers acquire knowledge than the learning of students who study textbooks and arguments. The latter kind of knowledge may be part of what needs to be done, but it is not sufficient for becoming a moral "expert" or for exercising moral creativity. Moral creativity, then, is something that needs to grow in experience. It is practical imagination that develops as one copes with problems. It is about skill, and skill requires training.

This means that moral problems cannot be solved once and for all by using moral principles. Rather, we have to learn moral creativity and grow in its use—without any guarantee that next time we can use the same solution. Dreyfus and Dreyfus (1980) propose an experimental and developmental view of ethics: ethical expertise requires skills and know-how as opposed to rules and propositional knowledge (know-that). Again, this does not mean that rules are superfluous; they are a good start. Dreyfus and Dreyfus (1980, 1991) make a distinction between different stages of moral knowledge, ranging from "novice" to "expert." Formalization and propositional knowledge are useful for novices, but moral maturity means that one has acquired an expertise that has grown in practice (see also Coeckelbergh, 2011). Morals experts are much more "creative" and "imaginative" in the sense that they no longer need rules or (other) explicit instruction. Their moral creativity and imagination have grown through trial and error and their knowledge is what Polanyi called "tacit knowledge" (Polanyi, 1967). Compare the acquisition of moral knowledge with learning to ride a bike or drive a car: first we need explicit instruction and rules, but someone who is an expert in this no longer needs the rules and can improvise, can be creative.

Furthermore, if moral creativity requires "a handling, using, and taking care of things which has its own kind of knowledge" (Crawford, 2009, p. 69; Crawford refers to Heidegger here), then this is not only good for others, but also contributes to our own virtue. For example, skilled work cultivates the virtue of "attentiveness" (Crawford, 2009, p. 82) and renders us involved "in a personal way": we care (Crawford, 2009, p. 95). Moral creativity could also be considered a virtue: it is the fruit of the development of moral skill. It is imagination: not the imagination of the detached Cartesian genius, but imagination-at-work. If this involves deliberation at all, it "is not disembodied cerebration deciding which action is derivable from ultimate principles, but is a form of engaged inquiry touched off by an uncertain situation" (Fesmire, 2003, p. 28). And insofar as it requires skill, it is more than "inquiry," a term that may still suggest a "mental" kind of imagination, unconnected to the social and material.

The previous analysis of moral creativity can be summarized by saying that moral creativity is a form of craftsmanship. It involves physical and bodily engagement with things; these physical, bodily practices give us tactile experience and relational understanding, a "tacit knowledge" of morality; and this produces virtue (*arete*). Thus, if moral creativity is a matter of craftsmanship, then the morally creative person is creative and imaginative in the way a cook is creative. A cook's knowledge is not merely theoretical and conceptual: his or her knowledge is tacit, has developed by means of tactile experience and involvement with food and people. Similarly, a moral "chef" is wise and creative: he or she has not the wisdom and imagination of Plato's statesman, but the wisdom and imagination of the moral cook, who has practical wisdom (*phronesis*) and exercises practical creativity and practical imagination. This enables him or her to respond adequately to the situation, the people, and the problem at hand.

The term phronesis is borrowed from Aristotle (2000), who uses it in his *Nicomachean Ethics*. According to him, practical wisdom is about perceiving what is good in concrete situations (Aristotle, 2000, 1142a) and about realizing a good that can only be realized, as Carr puts it, "in and through praxis itself" (Carr, 2006, p. 426). This form of knowledge can only be acquired in practice; one needs experience. Aristotle writes that "moral excellence comes about as a result of habit" (1103a) and that "the man who is to be good must be well trained and habituated" (1180a 14–15). He also gives an important role to law, deliberation, and reasoning (in this respect he is a forerunner of the modern moral philosopher), but his concept of practical wisdom has inspired many contemporary thinkers who are today categorized as "virtue ethicists." For example, MacIntyre has written that practical knowledge concerns "the capacity to judge and to do the right thing in the right place at the right time in the right way" (MacIntyre, 2007[1981], p. 150). This means that good is not pre-given; rather, it develops within a practice. Good craftsmanship seems to be good practice, with practice involving internal goods, goods internal to the practice. MacIntyre writes in *After Virtue*:

> By a "practice" I... mean any coherent and complex form of socially established cooperative human activity through which goods internal to practices are realized in the course of trying to achieve those standards of excellence which are appropriate to, and partially definitive of, that form of activity, with the result that human powers to achieve excellence, and human conceptions of the ends and goods involved, are systematically extended. (2007[1981], p. 187)

Similarly, we could say that moral creativity is achieved in the course of trying to be an excellent person, that it is not something that can be judged by external standards—ethical rules, laws, principles—but that it is part of moral expertise and moral craftsmanship, which no longer needs rules and explicit instruction and which is the result of a (long) process of moral development through engagement with people and with things. It involves not conceptual imagination, not the imagination of the (conceptual) designer, but practical imagination.

This kind of imagination is not only practical, but also *social*. In order to explain this we can again rely on Dewey's pragmatism, which understands morality as a social and imaginative enterprise. Again, we must question the modern-Romantic model of the hyper-individualistic genius, who acts as a single source of imagination and imprints his or her concept on the world. Instead, the social dimension of moral creativity compares well to the social dimension of drama and music. The moral life is not a solo music recital or a solo dance performance. Together we try to shape society and to cope with moral challenges. For example, coping with the current environmental problems cannot be done by an individual genius with a great concept or theory (for example, a scientist, artist, philosopher, politician), but requires pooled intelligence and action from all of us, with many people trying out different things.

This idea is in tune with Dewey's conception of moral deliberation as dramatic rehearsal. Fesmire uses the metaphor of jazz improvisation: when we play in a group of musicians we always respond to what others do (Fesmire, 2003) and indeed to what others have done in the past. There is a tradition:

A jazz musician...takes up the attitude of others by catching a cadence from the group's signals while anticipating the group's response to her own signals. Drawing on the resources of tradition, memory, and long exercise, she plays into the past tone to discover the possibilities for future tones. (Fesmire, 2003, p. 94)

We could also say that in dance the dancers respond to, and anticipate, what others do, and that the choreographer and the dancers do not create ex nihilo but dance into past movements to discover possibilities for future movements. Similarly, in morality we do not start from a blueprint, but from the habits and institutions that are handed over to us from the past. As in the arts, there is no creation ex nihilo. Even the painter starting from a blank canvas does not begin from nothing, but works within his or her connections with people and things, within his or her habits, within the tracks of personal history and societal history.

This is also an important dimension of the moral life and of moral creativity.

Moreover, since Aristotle, virtue is also understood to have an important social dimension: one can only develop into a virtuous person within a community. As Sennett emphasizes, the workshop of the craftsperson is a social space (2008, p. 73). The skilled work going in the workshop is also a form of working together and contributes to community building and to the growth of solidarity. Sennett writes that the medieval guild forged a strong sense of community, that it had rituals, and that its fraternities helped workers in need (2008, p. 60). Similarly, the moral craftsperson is not a solitary "moral worker" but always works with others. Moral improvement is a collaborative project, and in order to exercise moral creativity one has to work together with others. The moral imagination at work here is at the same time a social imagination.

Borgmann (1984) has also pointed to the social benefits of skilled work: skilled work is "bound up with social engagement" (1984, p. 42). He gives the example of what he calls "focal practices" such as gathering around a stove or drinking wine together. If there is an imagination that corresponds to skilled work, it is not the coining of an abstract concept but a very practical, embodied, and social activity. Similarly, the exercise of moral imagination is something people do together and cannot and should not be left to individuals in their study room. Moral imagination is needed where the moral challenges are: among people doing things together and gathering around the problems they face and in which they are involved. For example, we cannot detach ourselves from global problems: we all face them, in some way or other, and they can draw us together and demand a collaborative response from us.

This view of moral creativity also suggests a different, non-Platonic, and perhaps even non-modern view of how to change society: instead of making a blueprint of society and imagining the "perfect" society, the creativity we need here is one that imagines a society "on the move," a changing society as we find it, and a society of which we are *part* and in which we *participate*. It is not a society from which we are detached as philosopher-kings; we imagine better ways of doing things and of doing things together as part of a living society with its ongoing social and moral experiments.

But if society and morality are moving, this means we can no longer rely on fixed principles. As Dewey writes in *Human Nature and Conduct*, "life is a moving affair in which old moral truth ceases to apply" (1922, p. 164). Principles are part of our *toolkit*, but they are not more than that: they are not part of a transcendent collection of external truths,

truths that have nothing to do with our social practices. Instead, as I said when discussing the neo-Aristotelian view of virtue, good is internal to our practices. According to the Deweyan pragmatist view, principles are crystallized forms of moral experience, not the other way around as the Platonic view has it.

Furthermore, the relational view of moral subjectivity and moral creativity also has implications for thinking about the extent to which we can change the morality of our societies and of people. The modern-Romantic "genius" or Platonic model of moral creativity assumes that we can act as moral and social engineers in the sense of designing the concept of a perfect society with perfect people and then implementing this design, applying this concept to reality. But the Heideggerian, pragmatist, and neo-Aristotelian currents in contemporary philosophy suggest that this view of social change fails to acknowledge the existence of a moral tradition and fails to take into account the limits to moral change understood in terms of conceptual art and design.

If moral change depends on the cultural horizon that is already there and on the moral language and the material-technological structures and moral geographies that are already in place (Coeckelbergh, 2012), and if one's efforts to change morality crucially depend on what others do, then moral change is a slow, incremental process that cannot be completely controlled. In *Growing Moral Relations* (Coeckelbergh, 2012), I inquired into the conditions of possibility of moral change, and found that moral change depends on, and is limited by, our language, social relations, and culture, material and technological structures, moral-geographical patterns, spiritual and religious way of thinking and doing, and so on. Therefore, the metaphor of growth is a better one for describing moral change than, for example, design, production, or implementation. If morality is a relational matter and if moral creativity is deeply relational too, we cannot just change morality or society by having another concept. At most, we can do some gardening—preferably *creative* gardening in the practical and social sense articulated in this chapter.

Conclusion

I have argued that moral creativity is a matter of know-how and social engagement, not of designing moral principles or a Brave New World. We need imagination, but instead of only imagining an ideal moral world and its principles and rules, or using imagination *in* moral reasoning, we also need a more practical, social, and fluid kind of imagination, a creativity that is entangled with our experiences and our practices.

This kind of creative morality requires and promotes the moral development of humans rather than moral machines, and avoids a view of morality that confuses moral maturity with an approach to morality that may help moral beginners (children), but that when applied to adults is at best unfruitful and at worst pathological and dangerous. The kind of creativity and imagination we need in morality is not about designing or following rules and principles, but is about acquiring moral skill, attaining what I call "moral craftsmanship." Moreover, if changing the morality of one's life, practices, and society is not only a matter of "design," the metaphor of growth is more appropriate for describing moral change. This means that true moral creativity is about imagining new possibilities by trying out things in practice and by working together to make things better while drawing on the rich resources of moral experience—experience that has grown during our lifetimes and during the lifetimes of those who were struggling with moral and social challenges long before we were born. It is in this dance with contemporary, past, and future others that we have to imagine, discover, and create new moral possibilities.

References

Aristotle. (2000). *Nicomachean Ethics* (R. Crisp, trans.). Cambridge: Cambridge University Press.

Asimov, I. (1942). Runaround. *Astounding Science Fiction*, March, 94–103.

Bentham, J. (1970[1789]). *An Introduction to the Principles of Morals and Legislation* (J. H. Burns & H. L. A. Hart, eds.). London: Athlone Press.

Borgmann, A. (1984). *Technology and the Character of Contemporary Life: A Philosophical Inquiry*. Chicago, IL: University of Chicago Press.

Carr, W. (2006). Philosophy, methodology and action research. *Journal of Philosophy of Education, 40*(4): 421–435.

Coeckelbergh, M. (2007). *Imagination and Principles: An Essay on the Role of Imagination in Moral Reasoning*. Basingstoke: Palgrave Macmillan.

Coeckelbergh, M. (2011). Environmental virtue: Motivation, skill, and (in)formation technology. *Environmental Philosophy, 8*(2), 141–169.

Coeckelbergh, M. (2012). *Growing Moral Relations*. Basingstoke: Palgrave Macmillan.

Crawford, M. B. (2009). *Shop Class as Soulcraft: An Inquiry into the Value of Work*. New York: Penguin.

Dewey, J. (1922). *Human Nature and Conduct: An Introduction to Social Psychology*. London: Allen and Unwin.

Dreyfus, S. E., & Dreyfus, H. L. (1980). *A Five-Stage Model of the Mental Activities Involved in Direct Skill Acquisition*. Berkeley, CA: University of California Press.

Dreyfus, H., & Dreyfus, S. (1991). Towards a phenomenology of ethical expertise. *Human Studies, 14*(4), 229–250.

Fesmire, S. (2003). *John Dewey and Moral Imagination.* Bloomington: Indiana University Press.

Heidegger, M. (1996[1927]). *Being and Time: A Translation of Sein und Zeit* (J. Stambaugh trans.) Albany: State University of New York Press.

Johnson, M. (1993). *Moral Imagination: Implications of Cognitive Science for Ethics.* Chicago, IL: Chicago University Press.

MacIntyre, A. (2007[1981]). *After Virtue,* 3rd edn. Notre Dame, IN: University of Notre Dame Press.

Nagel, T. (1986). *The View from Nowhere.* Oxford: Oxford University Press.

Pappas, G. F. (2008). *John Dewey's Ethics: Democracy as Experience.* Bloomington: Indiana University Press.

Polanyi, M. (1967). *The Tacit Dimension.* New York: Anchor Books.

Popper, K. (2013[1945]). *The Open Society and Its Enemies.* Princeton, NJ: Princeton University Press.

Sennett, R. (2008). *The Craftsman.* New Haven, CT: Yale University Press.

Singer, P. (1975). *Animal Liberation.* New York: Random House.

Wallach, W., & Allen, C. (2009). *Moral Machines: Teaching Robots Right from Wrong.* Oxford: Oxford University Press.

3
Creativity in Ethical Reasoning

Robert J. Sternberg
University of Wyoming, USA

Two of the most creative and well-known studies ever done in psychology involved placing participants in ethically challenging situations. One experiment, originally conducted over a period of years in the 1960s by Stanley Milgram (see Milgram, 2010), asked participants to shock a "learner" in what was purported to be a verbal learning experiment. Unbeknown to the subjects, the shocks were fake and never delivered. The other experiment, conducted in 1971 by Philip Zimbardo (see Zimbardo, 2008), randomly divided subjects into guards and prisoners. In short order, the guards started acting like true prison guards with a sadistic streak, and the prisoners started acting like true prisoners cowed by those guards.

The behavior of the participants was ethically challenging as well as challenged. But the studies themselves were at least as challenging ethically as the participants' behavior because it was impossible fully to debrief participants. Whether they were "teachers" whose behavior might have killed a "learner" or "prison guards" who mistreated a "prisoner," simply telling participants that they were subjects in an experiment so their ethically challenged behavior was really all right does not work. The participants will have gone through their lives knowing that, given the opportunity, they acted in ways that by any reasonable standard were ethically unacceptable.

It is perhaps ironic that two of the most creative and widely cited studies ever conducted in psychology both involved ethical challenges, for the experimenters as well as the subjects. The challenge for the subjects, of course, was to act in an ethical way in responding to other subjects; the challenge to the experimenters was to figure out how to debrief subjects so that those subjects would not forever feel guilty after they had "learned" that they were not really ethical people. It is highly unlikely that either study, at least in its original form, would be approved by an ethics panel today. Clearly, creativity and ethics are not ready-made bedfellows. One easily can be creative without being ethical. In this chapter, though, I will argue that it is harder to be ethical without being creative.

Before continuing, it would be useful to define what I mean by both creativity and ethics. By creativity, I refer to thought and action that is novel to some degree and also compelling, or useful in some way. By ethics, I refer simply to doing the right thing. What is right depends on context, so ethical principles need to take into account the context of action.

Of course, we know that creativity and ethics often do not automatically go together. Creativity has a dark side (see Cropley et al., 2010; Sternberg, 2010a). In this chapter, I seek explicitly to address the creative aspects of ethical reasoning. The basic thesis is that ethical reasoning is difficult in part because it often requires a level of creative thinking that the individual doing the ethical reasoning lacks. More centrally, both ethical and creative action often require people to defy the crowd; that is, to do the opposite of what others are doing and expect one to do as well.

People—whether they follow the crowd or not—act unethically in many situations, even when the situations would not seem on the face of it to be particularly challenging at an ethical level. Sometimes this is because ethics means little or nothing to them. But more often, it is because it is difficult to translate theory into practice—in essence, to be creative in the ethical domain. Consider an example of this difficulty.

Latané and Darley (1970) opened up a new field of research on bystander intervention. They showed that, contrary to expectations, bystanders intervene when someone is in trouble only in very limited circumstances. For example, if they think that someone else might intervene, bystanders tend to stay out of the situation. Latané and Darley even showed that divinity students who were about to lecture on the parable of the good Samaritan were no more likely than other bystanders to help a person in distress who was in need of—a good Samaritan!

Gardner (1999) has wrestled with the question of whether there is some kind of existential or even spiritual intelligence that guides people through challenging life dilemmas. Coles (1998) is one of many who have argued for a moral intelligence in children as well as adults. Is there some kind of moral or spiritual intelligence in which some children are inherently superior to others? Piaget (1932) and Kohlberg (1984) believed that there are stages of moral reasoning, and that as children grow older they advance in these stages. Some will advance faster and further than others, creating individual differences in levels of moral development. Harkness, Edwards, and Super (1981) have questioned whether the stages are culturally generalizable. In contrast to the Kohlberg model, Gilligan (1982) argued that Kohlberg overly emphasized the development of principles of universal justice over a psychology of caring and compassion.

Some believe that ethical reasoning has a large non-rational component (for example, Rogerson et al., 2011), but the claim here is that ethical reasoning can be largely rational, although it usually is not because people fail to follow through on the complete set of steps needed to reach an ethical conclusion. And they often fail to follow through because they lack sufficient creative imagination to reach such a conclusion.

A model of ethical reasoning and its relation to creativity

Drawing in part on the Latané–Darley (1970) model of bystander intervention, I have constructed a model of ethical behavior that would seem to apply to a variety of ethical problems. The model specifies the specific skills that students need to reason and then behave ethically.

The basic premise of the model is that ethical behavior is far harder to display than one would expect simply on the basis of what we learn from our parents, school, and our religious training (Sternberg, 2009a, b, c). To intervene in an ethically challenging situation, individuals must go through a series of steps, and unless all of the steps are completed, the individuals are not likely to behave in an ethical way, regardless of the amount of training they have received in ethics, and regardless of their levels of other types of skills. The example I will draw on most is genocides, such as in Rwanda and Darfur, where there is the potential for outside intervention but the intervention in fact never happens, or happens only to a minor extent.

According to the proposed model, enacting ethical behavior is much harder than it would appear to be because it involves multiple, largely sequential steps. To behave ethically, the individual has to:

1. Recognize that there is an event to which to react.
2. Define the event as having an ethical dimension.
3. Decide that the ethical dimension is of sufficient significance to merit an ethics-guided response.
4. Take responsibility for generating an ethical solution to the problem.
5. Figure out what abstract ethical rule(s) might apply to the problem.

6. Decide how these abstract ethical rules actually apply to the problem so as to suggest a concrete solution.
7. Prepare for possible repercussions of having acted in what one considers an ethical manner.
8. Act.

Let us consider each step in turn.

1. Recognize that there is an event to which to react

In cases where there has been an ethical transgression, the transgressors often go out of their way to hide the fact that there is even an event to which to react. For example, many countries hide the deplorable condition of their political prisoners. The Nazis hid the existence of death camps and referred to Jews, Roma, and other peoples merely as being "resettled." The Rwandan government tried to cover up the massacre of the Tutsis and also of those Hutus who were perceived as sympathetic to the Tutsis. The goal of the transgressors is to obscure the fact that anything is going on that is even worth anyone's attention. One has to recognize that the situation as described by the government may be different from the actual situation. Put another way, one has to be creative in contemplating possibilities other than the one presented by those who wish to cover up their transgressions. People who are unwilling to defy the crowd—who instinctively conform to the will of others—are generally unwilling to take this step. It is for this reason that I often refer to creativity as a "decision" (Sternberg, 2003a, b).

Ivan Pavlov was creative, for example, in recognizing something others saw but failed to encode; namely, that dogs would salivate to the trainers who brought them food. Others saw the phenomenon, of course, but did not know what to make of it. Pavlov creatively recognized its importance.

When people hear their political, educational, or religious leaders talk, they may not believe there is any reason to question what they hear. After all, they are listening to authority figures. In this way, leaders, especially cynical and corrupt leaders, may lead their followers to accept corruption and even disappearances as non-events. It requires an extra creative step to consider other possibilities, and many people will not decide for creativity in this and other instances (see Sternberg, 2000).

2. Define the event as having an ethical dimension

Given that one acknowledges there is an event to which to pay attention, one still needs to define it as having an ethical dimension.

Given that perpetrators will go out of their way to define the situation otherwise—as a non-event, a civil war, an internal conflict that is no one else's business, and so on—one must actually redefine the situation to realize that an ethical component is involved. Redefinition of problem situations is one of the keys to creativity (Sternberg, 2000, 2003c). Again, a creative component is central to ethical reasoning.

In the case of the Nazi genocide, the campaign against Jews was defined as a justified campaign against an internal enemy bent on subversion of the state (Sternberg & Sternberg, 2008). It was, of course, not defined as genocide. To this day, the Turkish government defines the Armenian genocide as a conflict for which both sides must share the blame (Sternberg & Sternberg, 2008). And in Rwanda, the government defined the genocide as a fight against invading aggressors who came from outside the country and did not belong there in the first place. Redefining a situation requires creative effort, and most people simply do not decide for creativity (Sternberg & Lubart, 1995).

Creativity is in large part a function of defining situations in particular ways, whether in the ethical domain or any other. For example, Sir Alexander Fleming was not the first scientist to see mold destroy bacteria in a Petri dish. He was the first, however, to recognize that the mold was potentially valuable. It later became the basis for the antibiotic penicillin.

3. Decide that the ethical dimension is significant

If one observes a driver going one mile per hour over the speed limit on a highway, one is unlikely to become perturbed about the driver's unethical behavior, especially if that driver is oneself. Genocide is a far cry from driving one mile per hour over the speed limit. And yet, if one is being told by cynical, dishonest leaders that the events that are transpiring are the unfortunate kinds of events that happen in all countries—didn't America have its own Civil War?—then one may not realize that the event is much more serious than its perpetrators are alleging it to be. Again, if people are told that events have no significant ethical dimension—that they are routine events—then it takes an additional creative step on an individual's part to imagine otherwise. The example of Sir Alexander Fleming applies here as well. Fleming saw as important what other scientists had seen merely as a nuisance. Indeed, many scientific discoveries, as well as ethical ones, are based on someone seeing in a phenomenon what others do not see.

4. Take personal responsibility for generating an ethical solution to the problem

People may allow leaders to commit wretched acts, including genocide, because they figure it is the leaders' responsibility to determine the ethical dimensions of their actions. Is that not why they are leaders in the first place? Or people may assume that the leaders, especially if they are religious leaders, are in a uniquely good position to determine what is ethical. If a religious leader encourages someone to become a suicide bomber or to commit genocide, that "someone" may feel that being such a bomber must be ethical. Why else would a religious leader have suggested it?

Taking personal responsibility means redefining a situation as involving oneself in some way, not only others. And since it is so much easier to view an ethical dilemma as someone else's problem, many people do not make the creative step.

To some people, taking responsibility would scarcely seem to be a creative act. But this is exactly how awful things can go on in a country—Stalin's Russia, Hitler's Germany—without people stepping in actively to fight oppression. They simply do not take the extra creative step to realize that the situation is not just someone else's problem, but also their own. As German pastor Martin Niemöller is quoted as having said:

> First they came for the communists,
> and I didn't speak out because I wasn't a communist.
> Then they came for the socialists,
> and I didn't speak out because I wasn't a socialist.
> Then they came for the trade unionists,
> and I didn't speak out because I wasn't a trade unionist.
> Then they came for me,
> and there was no one left to speak for me.

> (http://en.wikipedia.org/wiki/First_they_came...)

5. Figure out what abstract ethical rule(s) might apply to the problem

Most of us have learned, in one way or another, ethical rules that we are supposed to apply to our lives. For example, we are supposed to be honest. But who among us can say we have not lied at some time, perhaps with the excuse that we were protecting someone else's feelings?

By doing so, we insulate ourselves from the effects of our behavior. Perhaps, we can argue, the principle that we should not hurt someone else's feelings takes precedence over not lying. Of course, as the lies grow larger, we can continue to use the same excuse.

When leaders encourage genocide, they clearly violate one of the Ten Commandments; namely, "Thou shalt not murder." This is why the killings, to the extent they are known, are posed by cynical leaders as "justifiable executions" rather than as murders. The individual must analyze the situation carefully to realize whether the term "murder" does in fact apply.

This step is primarily analytical, but analytical thinking is part of the creative process. Creative thinking is not merely original; it also is compelling or useful in some way. And in order to decide whether one's creative ideas are compelling or useful, one must analyze them before presenting them. That is, one must think analytically to ensure that the ideas are not merely new or original without being useful.

6. Decide how these abstract ethical rules actually apply to the problem so as to suggest a concrete solution

This kind of translation is, I believe, non-trivial. In our work on practical intelligence, some of which was summarized in Sternberg et al. (2000), we found that there is, at best, a modest correlation between the more academic and abstract aspects of intelligence and its more practical and concrete aspects. Both aspects, though, predicted behavior in everyday life. People may have skills that shine brightly in a classroom, but that they are unable to translate into real-world consequential behavior. This step, as applied to recognizing that murder is afoot in genocide, is primarily a function of the analytical part of creativity.

7. Prepare for possible repercussions of having acted in what one considers an ethical manner

When Harry Markopolos (see Markopolos, 2011) pointed out to regulators that Bernard Madoff's investment returns had to be fraudulent, no one wanted to listen. It was Markopolos who was branded as a problem, not Madoff. In general, when people blow the whistle, they need to be prepared for their bona fides to be questioned, not necessarily those of the person on whom they blew the whistle (as Marianne Gingrich discovered when she was branded a liar by her former husband, after her revelation that a decade earlier Newt Gingrich, a campaigner for family values, had wanted an open marriage and had been having an affair, later resulting in their divorce).

People think creatively when they imagine the possible repercussions of acting ethically—will they lose their friends, will they lose their job, will they lose their reputation? During the scandal involving the energy company, Enron, whistleblower Sherron Watkins lost all three. Similarly, when reports first came in of Nazi genocide, there was a general reaction of disbelief—how could such atrocities possibly be happening? Whistleblowers need to imagine all the things that could go wrong, but they also need to imagine what could go right and how they can maximize the chances of things going right. Such imagination requires creative thinking.

8. Act

In ethical reasoning, as in creativity, there may be a large gap between thought and action. Both often involve defying the crowd, and hence even people who believe a certain course of action to be correct may not follow through on it.

In Latané and Darley's (1970) work, the more bystanders there were, the less likely an individual was to take action to intervene. Why? Because one figured that, if something was really wrong, someone else among all the others witnessing the event would have taken responsibility. You are better off having a breakdown on a lonely country road than on a busy highway, because a driver passing by on the country road may feel that he or she is your only hope.

Sometimes, the problem is not that other people seem oblivious to the ethical implications of the situation, but that they actively encourage you to behave in ways you define as unethical. In the Rwandan genocides, Hutus were encouraged to hate Tutsis and to kill them, even if they were members of their own family (see discussion in Sternberg & Sternberg, 2008). Those who were not willing to participate in the massacres risked becoming victims themselves (Gourevitch, 1998). The same applied in Hitler's Germany. Those who tried to save Jews from concentration camps themselves risked being sent to such camps (Totten, Parsons, & Charny, 2004). It is easier to follow the crowd than to act creatively or, in many instances, ethically. And that is why corruption is so common throughout the world. Even when people know of it, they often reelect corrupt leaders, allowing the corruption to persist.

Teaching for ethical reasoning

We need to teach for ethical reasoning (Sternberg, 2010b). In recent years, we have seen the end of Bear Stearns, Lehman Brothers, Merrill

Lynch, and numerous other financial enterprises. Few people reached the depths of Bernard Madoff, the epitome of unethical behavior on Wall Street, who sits in a prison cell. The irony is that firms like Bear Stearns and Lehman Brothers hired only those they considered to be the best and the brightest. They recruited from the very top colleges and universities. It appears that whatever qualities one needs to be accepted by these institutions and to be graduated from them with distinction are not the qualities that would have led to success in these firms. In large part, university success reflects a student's ability to absorb a knowledge base and to reason analytically with it. However, success in business and in life require creative and ethical reasoning, none of which is at a premium in university life or in the standardized tests now used to admit students to universities. In a nutshell, we are selecting for and developing qualities that, while important, are woefully incomplete when it comes to success in the world.

The model applies not only to analyzing others but to evaluating one's own ethical reasoning. When confronted with a situation having a potential ethical dimension, students can learn literally to go through the steps of the model and ask how they apply to a given situation.

Effective teaching of ethical reasoning involves presenting case studies, but it is important that students also generate their own case studies from their own experience, and then apply the steps of the model to their own problems. They need to be actively involved in seeing how the steps of the model apply to their own individual problems. Most importantly, they need to think creatively as they use the model of ethical reasoning in thinking about ways of defining and redefining ethical dilemmas that enable them to get through the various steps.

As a university administrator, I, like other administrators, have discovered that students' ethical skills often are not up to the level of their ability test scores. Colleges run the full gamut of unethical behavior on the part of students: drunken rampages, cheating on tests, lying about reasons for papers turned in late, attacks on other students, questionable behavior on the athletic field. Faculty members, of course, are not immune either: few academic administrators probably leave their jobs without having had to deal with at least some cases of academic or other misconduct on the part of faculty. In hearing excuses students invent for work not done, I often have wished that students and faculty alike would apply their creativity to ethical rather than unethical uses.

Teaching a step-based model may seem to be self-contradictory to the notion of encouraging creativity. Where is the creativity if one recommends a fixed series of steps? Nevertheless, there is no contradiction.

The creativity is in the *use* of the steps. If one does not think creatively at each step of the model, one will not be able to use the model success-fully. The model is merely a framework. The creativity lies in the content one creates at each step. Teaching from such a model itself requires creativity. The creativity lies in assisting students to internalize the model—or some other model that potentially will lead them to ethical decisions. We already know that merely going to church or religious school in no way guarantees that one will apply the ethical principles one learns in these settings to one's life. The teacher creates transfer by creatively selecting and then deploying case studies that are relevant to students' own lives and then helping students to internalize the kind of thinking for which the model provides a framework.

In speaking of the challenges of leadership, and particularly of leaders who become foolish, I have spoken of the risk of ethical disengagement (Sternberg, 2008). Ethical disengagement (based on Bandura, 1999) is the dissociation of oneself from ethical values. One may believe that ethical values should apply to the actions of others, but one becomes disengaged from them as they apply to oneself. One may believe that one is above or beyond ethics, or simply not see its relevance to one's own life. Unless one seeks creatively to redefine the way one sees oneself, one sees oneself as ethical when in fact one has entered a period of downward ethical drift (Sternberg, 2012).

Schools should teach ethical reasoning; they should not necessar-ily teach ethics. There is a difference. Ethics is a set of principles for what constitutes right and wrong behavior. These principles are gener-ally taught in the home or through religious training in a special school or through learning in the course of one's life. It would be challenging to teach ethics in a secular school, because different religious and other groups have somewhat different ideas about what is right and wrong. There are, however, core values that are common to almost all these religions and ethical systems that schools do teach and reinforce: for example, reciprocity (the golden rule), honesty, sincerity, or compassion in the face of human suffering.

Ethical reasoning is how to think about issues of right or wrong. Processes of reasoning can be taught, and the school is an appropri-ate place to teach these processes. The reason is that, although parents and religious schools may teach ethics, they do not always teach ethical reasoning; or at least, not with great success. They may see their job as teaching right and wrong, but not how to reason with ethical principles. Moreover, they may not do as good a job of this as we would hope.

Is there any evidence that ethical reasoning can be taught with success? There have been successful endeavors with students of various ages. Paul (Paul & Elder, 2005), of the Foundation for Critical Thinking, has shown how principles of critical thinking can be applied specifically to ethical reasoning in young people. On the present view, for the instruction to be fully successful, teachers also would have to teach for creative thinking. DeHaan and his colleagues at Emory University have shown that it is possible successfully to teach ethical reasoning to high school students (DeHaan & Narayan, 2007). Myser of the University of Newcastle has shown ways specifically of teaching ethics to medical students (Myser, Kerridge, & Mitchell, 1995). Weber (1993) of Marquette University found that teaching ethical awareness and reasoning to business school students can improve through courses aimed at these topics, although the improvements are often short term. But Jordan (2007) found that as leaders ascend the hierarchy in their businesses, their tendency to define situations in ethical terms actually seems to decrease.

Ultimately, the greatest protection against ethical failure is wisdom—recognizing that, in the end, people benefit most when they act for the common good. Wisdom is the ultimate lifeboat (Sternberg, 2005; Sternberg, Jarvin, & Grigorenko, 2009; Sternberg & Jordan, 2005; Sternberg, Reznitskaya, & Jarvin, 2007). But for sure it requires creativity to figure out exactly what actions help to achieve a common good!

Conclusion

Deciding how to confront ethical challenges is one of the biggest dilemmas we will face in our lives (Sternberg, 2011a, b). However, when citizens fail and when leaders fail, it is not usually because they are not smart or knowledgeable enough. Rather, it is because they lack the creativity and ethical reasoning they need to get their businesses and their lives back on track.

References

Bandura, A. (1999). Moral disengagement in the perpetration of inhumanities. *Personality and Social Psychology Review, 3*, 193–209.

Coles, R. (1998). *The Moral Intelligence of Children: How to Raise a Moral Child.* New York: Plume.

Cropley, D. H., Cropley, A. J., Kaufman, J. C., & Runco, M. A. (eds.) (2010). *The Dark Side of Creativity.* New York: Cambridge University Press.

DeHaan, R., & Narayan, K. M. (eds.) (2007). *Education for Innovation.* Rotterdam: Sense Publishers.

Gardner, H. (1999). Are there additional intelligences? The case for naturalist, spiritual, and existential intelligences. In J. Kane (ed.), *Education, Information, and Transformation* (pp. 111–131). Upper Saddle River, NJ: Prentice-Hall.

Gilligan, C. (1982). In a different voice: Women's conceptions of self and morality. *Harvard Educational Review, 47*(4), 481–517.

Gourevitch, P. (1998). *We Wish to Inform You That Tomorrow We Will Be Killed with Our Families: Stories from Rwanda.* New York: Farrar, Straus & Giroux.

Harkness, S., Edwards, C. P., & Super, C. M. (1981). The claim to moral adequacy of a highest stage of moral judgment. *Developmental Psychology, 17*(5), 595–603.

Jordan, J. (2007). Taking the first step towards a moral action: An examination of moral sensitivity measurement across domains. *Journal of Genetic Psychology, 168*, 323–359.

Kohlberg, L. (1984). *The Psychology of Moral Development: The Nature and Validity of Moral Stages.* New York: HarperCollins.

Latané, B., & Darley, J. (1970). *The Unresponsive Bystander: Why Doesn't He Help?* Englewood Cliffs, NJ: Prentice-Hall.

Markopolos, H. (2011). *No One Would Listen: A True Financial Thriller.* New York: Wiley.

Milgram, S. (2010). *Obedience to Authority: An Experimental View.* New York: Harper Perennial.

Myser, C., Kerridge, I. H., & Mitchell, K. R. (1995). Teaching clinical ethics as a professional skill: Bridging the gap between knowledge about ethics and its use in clinical practice. *Journal of Medical Ethics, 21*(2), 97–103.

Paul, R., & Elder, L. (2005). *Critical Thinking: Tools for Taking Charge of Your Learning and Your Life,* (2nd edn.). Upper Saddle River, NJ: Prentice-Hall.

Piaget, J. (1932). *The Moral Judgment of the Child.* London: Kegan Paul, Trench, Trubner & Co.

Rogerson, M. D., Gottlieb, M. C., Handelsman, M. M., Knapp, S., & Younggren, J. (2011). Nonrational processes in ethical decision making. *American Psychologist, 66*(7), 614–623.

Sternberg, R. J. (2000). Creativity is a decision. In A. L. Costa (ed.), *Teaching for Intelligence II* (pp. 85–106). Arlington Heights, IL: Skylight Training and Publishing.

Sternberg, R. J. (2003a). Creativity is a decision. *APA Monitor on Psychology, 34*(10), 5, 54.

Sternberg, R. J. (2003b). The development of creativity as a decision-making process. In R. K. Sawyer, V. John-Steiner, S. Moran, R. J. Sternberg, D. H. Feldman, J. Nakamura, & M. Csikszentmihalyi (eds.), *Creativity and Development* (pp. 91–138). New York: Oxford University Press.

Sternberg, R. J. (2003c). *Wisdom, Intelligence, and Creativity, Synthesized.* New York: Cambridge University Press.

Sternberg, R. J. (2005). A model of educational leadership: Wisdom, intelligence, and creativity synthesized. *International Journal of Leadership in Education: Theory & Practice, 8*, 347–364.

Sternberg, R. J. (2008). The WICS approach to leadership: Stories of leadership and the structures and processes that support them. *Leadership Quarterly, 19*(3), 360–371.

Sternberg, R. J. (2009a). Ethics and giftedness. *High Ability Studies, 20*, 121–130.

Sternberg, R. J. (2009b). A new model for teaching ethical behavior. *Chronicle of Higher Education, 55*(33), April 24, B14–B15.

Sternberg, R. J. (2009c). Reflections on ethical leadership. In D. Ambrose & T. Cross (eds.), *Morality, Ethics, and Gifted Minds* (pp. 19–28). New York: Springer.

Sternberg, R. J. (2010a). The dark side of creativity and how to combat it. In D. H. Cropley, A. J. Cropley, J. C. Kaufman, & M. A. Runco (eds.), *The Dark Side of Creativity* (pp. 316–328). New York: Cambridge University Press.

Sternberg, R. J. (2010b). Teaching for ethical reasoning in liberal education. *Liberal Education, 96*(3), 32–37.

Sternberg, R. J. (2011a). Ethics: From thought to action. *Educational Leadership, 68*(6), 34–39.

Sternberg, R. J. (2011b). Slip-sliding away, down the ethical slope. *Chronicle of Higher Education, 57*(19), A23.

Sternberg, R. J. (2012). Ethical drift. *Liberal Education, 98*(3), 60.

Sternberg, R. J., & Jordan, J. (eds.) (2005). *Handbook of Wisdom: Psychological Perspectives*. New York: Cambridge University Press.

Sternberg, R. J., & Lubart, T. I. (1995). *Defying the Crowd: Cultivating Creativity in a Culture of Conformity*. New York: Free Press.

Sternberg, R. J., & Sternberg, K. (2008). *The Nature of Hate*. New York: Cambridge University Press.

Sternberg, R. J., Jarvin, L., & Grigorenko, E. L. (2009). *Teaching for Wisdom, Intelligence, Creativity, and Success*. Thousand Oaks, CA: Corwin.

Sternberg, R. J., Reznitskaya, A. & Jarvin, L. (2007). Teaching for wisdom: What matters is not just what students know, but how they use it. *London Review of Education, 5*(2), 143–158.

Sternberg, R. J., Forsythe, G. B., Hedlund, J., Horvath, J., Snook, S., Williams, W. M., Wagner, R. K., & Grigorenko, E. L. (2000). *Practical Intelligence in Everyday Life*. New York: Cambridge University Press.

Totten, S., Parsons, W. S., & Charny, I. W. (eds.). (2004). *Century of Genocide: Critical Essays and Eyewitness Accounts*. London: Routledge.

Weber, J. (1993). Exploring the relationship between personal values and moral reasoning. *Human Relations, 46*, 435–463.

Zimbardo, P. (2008). *The Lucifer Effect*. New York: Random House.

4
Moral Creativity and Creative Morality

Qin Li and Mihaly Csikszentmihalyi
Claremont Graduate University, USA

> In the interest of doing my job as quickly and efficiently as possible, I want to introduce myself now so we don't have to waste valuable time later with idle chit chat. I am Dr Gregory House ... I am a board certified diagnostician with a double specialty of infectious disease and nephrology. I am also the only doctor currently employed at this clinic who is forced to be here against his will ... but you don't need to worry because for most of you, this job could be done by a monkey with a bottle of Motrin. Speaking of which, if you're particularly annoying, you may see me reach for this. This is Vicodin. It's mine ... I do not have a pain management problem. I have a pain problem. But who knows, maybe I'm wrong. Maybe I'm too stoned to tell.
>
> David & Singer (2004)

The popular image of the amorality of the creative person

Gregory House is a misanthropic drug-addicted genius diagnostician in the Fox network series *House, M.D.* House exemplifies a popular notion that creative individuals are conceited people concerned only with their creative accomplishment and held little concern for others. In fact, the evidence to support this popular image is not difficult to find. There have been many artists and scientists in the past who have claimed that their work cannot be judged by the standards of conventional morality, and have backed up their words with their actions. From these examples many people have inferred that somehow morality is incompatible with creativity, and that creative individuals arrogantly hold themselves above the rest of humanity, ignoring and disdaining the values that keep the community together. In some cases, people draw the inference that ethical concerns can be detrimental to creativity.

In reality, however, the situation is very different and much more complex. This chapter, based on interviews with creative individuals and on historical examples, describes the moral perceptions of creative individuals; specifically, their ethical views of their work and of society. It suggests a way of understanding how moral issues are addressed by creative individuals, through the concept of "creative morality" reminiscent of Aristotle's notion of phronesis. Phronesis, or "practical wisdom," refers to decisions made based on practicality as well as universal values. Aristotle adds that phronesis requires experience, which comes with age.

The data used in this chapter derives from a qualitative study of the lives of creative people directed by Csikszentmihalyi from 1990 to 1995 (cf. Csikszentmihalyi, 1996). By focusing on individuals over 60 years old, the study ensured that participants had demonstrated that their creativity was not due to one single lucky idea or event, but was supported by and exemplified a lifetime of accomplishments. Because the original interviews did not specifically target the issue of ethics and morality, here we will use only interviews with individuals who touched on that topic. Quotes not identified otherwise come from the Creativity in Later Life Study; quotes from other sources are cited accordingly.

A number of the interviewees were Nobel Prize recipients and many others received equally esteemed awards or recognition for their achievements and contributions to their domains. Thus the sample represents Big-C creativity, or creativity that changes the world in some way. We believe that by understanding the ethical and moral outlooks of individuals who are indisputably creative, we will have a better chance of understanding the connection between creativity and morality.

Why is the relationship between morality and creativity relevant in the first place? It is important because of the repercussions that Big-C creativity has on the world. Big-C creativity is not a personal phenomenon; it is the product of a social and cultural context. In turn, it also changes the culture and the society in some way. Given the wide impact of Big-C creativity, it matters whether its practices are immoral, amoral, or follow some moral laws. But what is morality?

There are two main perspectives on morality. The first, conventional morality, refers to "morality as an existing institution of a particular society." The second, universal morality, "is that of morality as a universal idea grounded in reason" (Deigh, 2010, p. 8). Conventional morality is a set of norms endorsed by society, whereas universal morality is supposedly shared by humanity as a whole. A third form of morality is represented by professional *ethics*—-a term that originally referred to the habits peculiar to a community or group (Deigh, 2010). In this chapter, professional ethics refers to the codes shared and expected of specific professional groups. One can view morality as applying to all members of a society, whereas ethics is career specific.

Many of the individuals we interviewed appear to fit into what the philosopher Ogilvy (1976) called "ethical anarchists." There are two extreme positions in ethics. The ethical naturalist sees universal principles or generalizations as law. Just as laws of nature govern the natural world, generalizations of morality govern moral decisions. At the other extreme, the ethical anarchist views these same generalizations as guidelines, which may be invalid in particular circumstances. Ogilvy explains:

> [b]ecause the uniqueness of each individual is the starting point of his investigations, he recognizes that a particular situation may render a plausible generalization irrelevant. To cite a very simple example, "honesty is the best policy" will seem irrelevant when a mad man asks where the ammunition is stored. (Ogilvy, 1976, p. 5)

Unlike the laws of physics, moral decisions can be messy because of conflicting choices and dilemmas. Every ethical decision is unique because of the uniqueness of the individual and the context. Each situation will also exhibit shared universals, such as belief systems, customs, or knowledge. Although the ethical anarchists favor the individual particulars, they are also aware of—but do not bind to—the generalizations. Few situations are black and white; when genuine conflict occurs the ethical anarchists are more likely to consider the circumstances, the particulars, and the generalizability of a situation to make sound ethical judgments (Ogilvy, 1976).

Prior qualitative analysis of the Creativity in Later Life Study had already illuminated some possible findings on creative individuals and their morality. Although one would assume that their demanding workload leaves them little time for other concerns, surprisingly, one third of the interviewees talked about duty and obligations beyond their work (Choe, Nakamura, & Csikszentmihalyi, 1996). Choe et al. (1996) found that of the 18 different values spontaneously mentioned in the interviews, concern for social wellbeing stood out in every instance as the ultimate value.

Across domains, Choe et al. (1996) found that the interviewees mentioned the priority of doing quality work: "doing careful, sound, exacting work is a principle which governs their efforts ongoing [sic]."

Performing good-quality work is the core, if not most important, principle of professional ethics. Between domains, Choe et al. (1996) found that scientists tend to talk more about fortitude/perseverance and hard work/diligence than do artists. Conversely, artists speak more about learning and experiencing, which is a natural outgrowth of art.

After reflecting on the different types of morality mentioned and the possible actions taken, we constructed a matrix of morality and their applications (Table 4.1). The three types of morality with three possible actions each (denied, ignored, and observed) generated 27 possible combinations: $3*3*3 = 27$. The interviewees will be classified using the matrix as a way to discuss the different creative individuals' actions in different moral dimensions. The model in Table 4.1 is extremely simplified. In the first place, it does not indicate how strongly a person observes or denies the ethical or moral standards. Nor does it indicate how widely the particular position applies: some people may take one kind of moral stance in one case, but another moral stance in another case, depending on the situation. Nevertheless, we might try to apply the model to one event at a time, to learn how it works in specific situations. Now, we shall turn to the interviews we collected, and examine some selected illustrative case studies.

Morality and science: A case of moral dilemma in science

At 8:15 a.m. on August 6, 1945, the 393d Bombardment Squadron B-29 dropped the "Little Boy" over Hiroshima, Japan; this first atomic bomb left total destruction within one mile's radius of its detonation point, with further damage reaching miles away. It instantly killed 70,000 people and more would die from exposure.

Back in the United States after the nuclear attack, physicists emerged as heroes who made possible Japan's surrender and the eventual end of the Second World War. In the following years, the Los Alamos division

Table 4.1 A model of dimensions of morality applied to the conduct of creative individuals

		Actions		
Moral dimensions	Conventional morality	Denied $= -1$	Ignored $= 0$	Observed $= 1$
	Universal morality			
	Professional ethics			

of the Manhattan Project would take the most vocal stand against future nuclear warfare, with division director J. Robert Oppenheimer and theoretical division head Hans Bethe as lead voices. Why did these physicists choose to take a stand against the very thing that brought victory to the United States?

Hans Bethe was one of the interviewees for the Creativity in Later Life Study. In a collection of essays, he stood his ground on the belief that regardless of the consequences, knowledge should be pursued for its own sake; but emphasized that scientists bear the responsibility for their contributions. Bethe observed conventional morality by acknowledging a sense of personal responsibility, and observed professional ethics by pursuing knowledge for science's sake. Universal morality was not considered:

> Should scientists refrain from working on dangerous problems? In my opinion, it would be impossible to have all competent scientists on all sides of a conflict agree to cease working on arms research ... whether or not their government responds to their advice [about the risks], scientists have an obligation to speak out publicly when they feel there are dangers ahead in any planned weapon development. (Bethe, 1991, pp. xv–xvi)

There is a distinction between knowledge and its application: knowledge is an inherent good in science, whereas whether its applications turn out good or bad is circumstantial (Schweber, 2000). The essence of science is the pursuit and establishment of knowledge (Sample, 2009). Knowledge, however, often comes at a price. Bethe and the Los Alamos physicists believed that pure knowledge is good, even if the knowledge can lead to potential evil.

Bethe stated the moral dilemma that the physicists involved in the Manhattan Project felt at the time: "most of the physicists who have had contact with atomic weapons are deeply disturbed. Most of them are very sensitive to moral values. At the same time, it is difficult for them to know what to do" (Bethe, 1991, p. 177). The scientists working at Los Alamos did what they had to do because they believed they were helping to end the war. They believed they were helping to save Western democracy, hoping that the development of the bomb would ensure victory and peace. Bethe said as a scientist that his service was obligatory, but that did not quiet the conflict to his own moral beliefs about not utilizing the bomb. Bethe also did not believe, within the context of the war, that they could refuse to work on nuclear weapons because that "would set up the scientific community as a superpolitical body" (Schweber, 2000, p. 170).

When the United States was moving on to the development of the H-bomb, Bethe vehemently protested that "its use would be the betrayal of all standards of morality and of Christian civilization itself...there can be only one justification for our development of the hydrogen bomb and that is to prevent its use" (Schweber, 2000, p. 161). If the dimensions of morality were arranged hierarchically, universal morality would encompass both conventional and professional ethics. In terms of Bethe's position for the H-bomb, he advised against using the bomb for the sake of a universal morality—to refrain from wiping out humanity:

> I believe that we would lose far more than our lives in a war fought with hydrogen bombs, that we would in fact lose all our liberties and human value at the same time, and so thoroughly that we would not recover them for an unforeseeably long time. (cited in Schweber, 2000, p. 162)

Initially, Bethe worked on ideas far removed from possible application and denied the moral responsibility of science. He took the conventional morality perspective in which morality is embedded in society: "science has nothing to say or to contribute to these human values, for or against them. Nor can science tell us which direction we should proceed to establish such values" (Schweber, 2000, p. 15). After the atomic bomb, Bethe sought to bridge the separation between science, morality, and even aesthetic. He advocated that growth in knowledge via science will lead people to make rational choices for the good of humanity—a universal morality perspective.

Despite speaking out against the H-bomb, he joined its project team after the outbreak of the Korean War: "If I didn't work on the bomb somebody else would—and I had the thought that if I were around Los Alamos I might still be a force for disbarment" (Schweber, 2000, p. 166). Despite Bethe's strong opposition to future nuclear weapons and warfare, he was a supporter of nuclear energy, arguing publicly for its necessity. As Schweber noted, "to this day, Bethe believes that the benefits that will accrue from the peaceful uses of atomic energy will justify the unraveling of nuclear power" (Schweber, 2000, p. 22).

Bethe tried to bring all three aspects of morality into alignment and at times he changed his emphasis from one dimension to another. His struggle to reconcile the ethics of science with universal values is a good illustration of how difficult it is for a creative scientist to find a stable moral stance.

Morality is often thought of as human-to-human conduct, but biologist Edward O. Wilson extended it to a species-to-species perspective. Wilson is perhaps a rarity in that his work as a biologist has led him to be characterized as observing universal and professional morality while ostracizing himself by rejecting conventional morality. A Google search for Wilson will pull up his biodiversity foundation, an organization he started in his 50s, where his motto is to "preserve every scrap of biodiversity as priceless while we learn to use it and come to understand what it means to humanity." This is a commitment to educate others about the need for biodiversity and to conserve all life on earth. In his book *The Diversity of Life* (1992), he coined the term "sixth-extinction" to describe the human destruction of biodiversity, which he argued will ultimately lead to the demise of humanity (as cited in Wilson, 2006).

Wilson has had an illustrious writing career with over 25 books and two Pulitzer prizes. During the interview for the Creativity in Late Life Study, he said, "scientists have to write books now if they want to reach a large audience quickly and not wait for the trickle down effect... of a seminal work." Wilson believes in the obligation of scientists to engage actively and educate the public on important scientific discoveries. Yet, at the same time, some of his most successful books have also been the most controversial. *Sociobiology* (1975) created a great deal of uproar in the scientific community and garnered as much praise as criticism. The objection raised to the book regarded it as genetic determinism:

> Mine was an exceptionally strong hereditarian position for the 1970s. It helped to revive the long-standing nature–nurture debate at a time when nurture had seemingly won. The social sciences were being built upon that victory... Many critics saw this challenge from the natural sciences as not just intellectually flawed but morally wrong. If human nature is rooted in heredity, they suggested, then some forms of social behavior are probably intractable. (Wilson, 2006, pp. 334–335)

Although *Sociobiology* was well received by biologists, the book was questioned and attacked by social scientists and Marxists on ethical and moral grounds. Some felt so strongly against it that a group of university faculty, scientists, teachers, doctors, and students jointly wrote a letter to the editor of the *New York Review of Books* in response to a review of Wilson's book. The letter was an attempt to refute Wilson's academic credibility and drew a parallel between his viewpoint and eugenics. Perhaps the most dramatic scene in the *Sociobiology* controversy was when protestors overtook the stage at the 1978 meeting of the

American Association for the Advancement of Science (AAAS), where Wilson was to deliver a speech.

During our interview, Wilson noted, "I went from being one of the most politically incorrect scientists in the country, to being politically one of the most correct. It is fascinating to be pilloried by the left in the 70s and now not pilloried, but mentioned with some distaste by the far right." He was reflecting on a change in conventional morality: in the 1990s the United States became more accepting of the position on the large role of genes on behavior that Wilson was advocating in the 1970s. For much of Wilson's career, he had ignored or denied conventional morality, and persisted in propounding a universal morality and professional ethics. Luckily for him, conventional morality eventually changed to favor his view, but for others that may never happen.

As the Nobel Laureate Paul Samuelson said of Wilson's work on evolution and human behavior, "this subject is an intellectual and doctrinal minefield"; one may never know where the bomb is buried and when it will explode. Wilson went through quite an ordeal in order to keep his academic position. He was taken aback by the line of attack and admitted that the controversy arose partly because of his naïveté about Marxism and the political far left. The lack of support from his colleagues further exacerbated his lack of confidence. However, his belief in his scientific training helped him keep his footing:

Then I rethought my own evidence and logic. What I had said was defensible as science. The attack on it was political, not evidential. [They] had no interest in the subject beyond discrediting it. They appeared to understand very little of its real substance... In a few more weeks, anger in turn subsided and my old confidence returned, then a fresh surge of ambition. (Wilson, 2006, p. 339)

Without a doubt, there are consequences to taking the maverick position on a subject. However, Wilson persevered and despite his setbacks has become an important figure in science. The earlier controversy showed his deep commitment to professional ethics and his denial of conventional morals. In his later years, Wilson focused more on universal morality, which was also deeply intertwined with his professional ethics.

Scientists not only educate and mentor the general public, they also do the same for their mentees. Rosalyn Yalow emphasized the importance of mentorship and nurturing the future generation. In her acceptance speech for the 1977 Nobel Prize in Physiology or Medicine for her and Solomon A. Berson's work on radioimmunology, Yalow spoke directly to the students present when she said:

> If we are to have faith that mankind will survive and thrive on the face of the earth, we must believe that each succeeding generation will be wiser than its progenitors. We transmit to you, the next generation, the total sum of our knowledge. Yours is the responsibility to use it, add to it, and transmit it to your children...We bequeath to you, the next generation, our knowledge but also our problems. ("Rosalyn Yalow–Banquet speech," 1977, para. 4)

Her speech was an observance of universal morality because of her concern for the survival of future generations and humankind itself. Yalow also expressed the generativity in her laboratory when describing the professional ethic that she upholds:

> It is caring about people and caring about their development...We always kept a small laboratory and we were always concerned with the development of the people who came into the laboratory. They were not an extra pair of hands for us; they were there to learn how to be their own pair of hands.

Yalow saw the lab workers as "professional children," in whose success she took pride. Moreover, she stressed that in her laboratory, not merely the research technique was important but also her and Berson's philosophy: "I have never aspired to have, nor do I want, a laboratory or cadre of investigators-in-training which is more extensive than I can personally interact with and supervise." In addition, she also made sure that she was a model for ethical behavior where "they would see that we did it, so they would do it." Her professional obligations also included educating other people: "[what] I am concerned [with] nowadays is educating people in areas in which I think are badly undereducated...I lecture at colleges, I lecture to the public, I lecture even to other scientists."

Considering Yalow's position as a female in a field dominated by men, she talked openly about the importance of empowering women, which was an active rejection of social norms at that time. It may not be a stretch to say that it would be amoral to deny the other half of humanity a chance at equal opportunity:

> The failure of women to have reached positions of leadership has been due in large part to social and professional discrimination...we

must feel a personal responsibility to ease the path for those [women] who come afterwards. The world cannot afford the loss of the talents of half its people if we are to solve the many problems which beset us. ("Rosalyn Yalow–Banquet speech," 1977, para. 3)

Yalow's success as a female Noble Prize–winning scientist overturned the conventional morality of her time. She observed universal morality through her advocacy of the survival and livelihood of future genera-tions, and also observed professional ethics by the way she managed her laboratory.

The role of scientist as educator resonates through all three inter-views. It appears to be a common theme among the academic elite that they need to educate the public and disseminate important knowledge. As predicted, the three scientists had a high regard for good work and a sense of social responsibility. For Bethe and Wilson, social concern was a direct extension of their scientific career, whereas for Yalow, it was of a more generative nature. A pattern that seems to emerge is the shift and change in moral dimensions: observance of professional ethics came first for all three, yet the focus on professional ethics had inevitably led them to accept, deny, or at times change conventional morality and later to strive for universal morality.

Art and morality: A case of apathy among artists

Oscar Wilde prefaced the 1891 edition of *The Picture of Dorian Gray* as follows:

> The artist is the creator of beautiful things. To reveal art and con-ceal the artist is art's aim... There is no such thing as a moral or an immoral book. Books are well written, or badly written... the morality of art consists in the perfect use of an imperfect medium. No artist desires to prove anything... No artist has ethical sympa-thies... Thought and language are to the artist instruments of an art... The only excuse for making a useless thing is that one admires it intensely.

The preface was written in response to the critics and public who deemed *The Picture of Dorian Gray* amoral primarily for its depiction of homosexuality. It elucidated Wilde's perspective on morality and art: he claimed that art should not be judged on moral standards. He argued that art is made only for aesthetic and not for moral judgments. Thus, any moral lessons received from the art reflect the audience, not the art. Moreover, art does not have a function, and its uselessness is only justified by its aesthetics; it is art for art's sake. Wilde's perspective on art observed professional ethics while ignoring conventional and universal morality. What is art? Art's Latin and Greek origins—*ars* and *techne*, respectively—mean human skills or artisanal skills. Art (of all genres) as we know it today is a modern idea dating back to the eighteenth century. Prior to that, art was an inclusive system of decoration for objects or activities. The split of art into functional and fine art developed in post-Renaissance Europe. The separation led to showcase venues (for example, theaters, museums, galleries) that further elevated the status of fine art by shedding any last bit of function and turning it into an object of pure aesthetics (Shiner, 2001).

The debate about "art for art's sake" and the "social responsibility of art" is a recent one, made possible only after fine art obtained its status as an autonomous domain (Shiner, 2001, p. 222). Up until the first half of the nineteenth century, most people believed that art should be didactic; that is, that art should entertain, teach, and closely reflect reality (Marangoz, 2006). The political and social unrest of nineteenth-century Europe left some artists seeking shelter in fine art, an autonomous realm separate from "police power and middle-class materialism"; simultaneously, it created hopefuls who saw art as a "social instrument" of change (Shiner, 2001, p. 223). Regardless of which position one takes in the debate, both parties recognize art as a separate realm from, but also with some relation to, society.

If Wilde were to exemplify the role of an amoral writer who disregarded conventional and universal morality, Madeleine L'Engle might be characterized as a moral writer. L'Engle is known for her prize-winning novel *A Wrinkle in Time* (1962), which dealt with the problems of evil, love, and moral purposes. L'Engle once said: "Why does anybody tell a story? ... [It is because of the] faith that the universe has meaning, that our little human lives are not irrelevant, that what we choose or say or do matters" (Martin, 2007). Observing professional ethics and universal morality, she commented:

> I am responsible for what I write; I know that I influence a lot of people. And therefore, I have to have a certain concern that what I write is not destructive ... I think you have to get them out with some kind of hope. I don't like hopeless books ... I don't think I could write hopeless books ... books that make you think life is not worth living.

L'Engle's awareness that she, as a writer, had a wide influence prompted her to write with a sense of keen responsibility to her audience. Simultaneously, her leaving hope for the readers spoke to a universal kindness. At the time of the interview, L'Engle was working on another novel about human complexity, dealing with the issues and struggles people face in life:

> Life is not supposed to be what the television commercials promise it's supposed to be . . . easy and comfortable and nothing's ever going to go wrong as long as you buy the right product, and if you have the right insurance everything is going to be fine . . . Terrible things happen and those are the things that we learn from.

L'Engle believed that books should be a source of inspiration and hope to a life already full of adversity. A novel can become a conversation between the author and the readers, such that when the readers close the book, they have the heart to go forward and face the world. Whereas Wilde detached himself from his work, L'Engle said there was no objectivity in writing: "everything anybody writes is based on personal experience, no matter how much they may deny it . . . I can't write about a place that I haven't known with all five senses."

In response to attempts at banning her book and criticism against her, L'Engle said, "It's very ironic to me that these censors are all people who call themselves Christians, and they're trying to define this little tiny god, from wicked people like me who are threatening to God. Uh, the sad thing is that they're getting books off the shelves." Instead of focusing on the personal attacks on her work, she felt that it was unjust to deprive readers, and more importantly children, of certain books because of a subgroup's imposition of its own values. L'Engle observed and intimately tied professional ethics to universal morality, but ignored conventional morality.

Another writer, Nadine Gordimer, was awarded the 1991 Nobel Prize in Literature, with the citation "who through her magnificent epic writing has—in the words of Alfred Nobel—been of very great benefit to humanity" ("Nadine Gordimer–Facts," 1991, para. 3). In a 2005 interview with the Nobel Prize organization, Gordimer said: "The purpose of writing [is] merely to explain the mystery of life, and the mystery of life, includes of course, the personal, the political" (Stanford & Gordimer, 2005). Writing is an exploration of life, self, and the world. Similar to L'Engle's comment that writers draw from personal experience, Gordimer admitted the impact of the social environment on her work—at times writing became a way of resolving personal crises. Like Wilson, Gordimer denied conventional morality and was working, both indirectly through her novels and directly via political involvement, to challenge convention even if it meant facing her fears. However, maintaining her professional ethics, the "aesthetic and artistic integrity" of a writer, was her priority:

> I was actually petrified... But then there came the absolutely irresistible political pressure, [I] couldn't go on living inside Africa, as a White South African, articulate... without speaking out about what was happening there. So I had to... I was not going to become a propagandist in my writing, my writing... has to have its aesthetic and artistic integrity... I had to take civic responsibility.

During the interview, Gordimer said that in sifting through some of her old stories for a new book collection, she noticed that she was obsessed with the theme of betrayal from the end of the 1970s through the 1980s. Her books often reflected the events surrounding her, whether it was growing up in apartheid South Africa or being an anti-apartheid activist. Her context also drove her to speak out against the chaos boiling in her country. She became political by choice and made a conscious decision about the type of message her work would convey.

Gordimer saw her ability to write as a gift, but in order to write, one must live a full life; a writer needs experiences from which to draw inspiration. To write is a solitary activity, yet the required writing material must come from having lived fully. As Gordimer said, "from this paradoxical inner solitude our writing is what Roland Barthes called 'the essential gesture' towards the people among whom we live, and to the world; it is the hand held out with the best we have to give" ("Nadine Gordimer–Banquet speech," 1991, para. 4).

To write the best one can as a way of gifting the world follows a high professional ethical calling. When facing social injustice, the phrase "action speaks louder than words" may not be well suited to writers. Gordimer believed that a writer's pen was her greatest weapon, a "non-action" action. She witnessed a time at which writers were suppressed by repressive regimes, and considered that it is of utmost importance for writers to resist and be of service to humanity by being truthful; their words "never changed by lies, by semantic sophistry, by the dirtying of the word for the purposes of racism, sexism, prejudice, domination, and the glorification of destruction" ("Nadine Gordimer–Nobel lecture: Writing and being," 1991, para. 22). Again, Gordimer exemplifies a union of universal morality with professional ethics while fighting to change the conventional morality.

Painters express similar sentiments about morality as writers. Among our interviewees, Ellen Lanyon was one who believed that the visual arts are not isolated from reality and should be didactic. When advising younger artists, she suggested that they should be well rounded and engaged in the world. More precisely, young people should be "aware of what is going on in the world and be interested in current history. And be a person of your generation. And also have an attitude that what you're doing is somehow a contribution to your culture." Unlike Wilde's declaration of art as useless aside from stimulating pleasure, Lanyon repeatedly stressed its importance as being impactful on the world. Like Gordimer, Lanyon believed her talent to be a gift she had to offer to the world. And like Yalow and Gordimer, she connected professional ethics with universal morality. In her words:

> I've never been an ivory tower painter . . . I don't believe that you paint for yourself. I think that you paint not to sell, but to give . . . You have a special vision, and this vision compounds itself with other artists' vision, and you create the social order. You create the culture that develops through history. You take from the past, and you give to the future.

Lanyon was indicating that one does not create for selfish reasons but rather to contribute to the culture, generation, and history of the world, and to do so because one is capable of it. Her advice to young artists was to make sure that they contributed to the world. It was also necessary for her to make a conscious attempt at making an impact on the world; there was an intentional purpose to her work. She wanted the audience to be engaged and be provoked by her paintings. To involve them, she said:

> When you look at one of these images, you are going to start to really think about what's going on. Now you may not get the same message that I'm trying to put out, but you're going to get something because there's a provocative element there.

Nor did Lanyon see art as detached creations: her art was a reflection of herself, thus, when she grew, so did her work; her work in the 1990s was different from that in the 1940s. In speaking about the differences, she said:

Now there are a lot of people who'll say, "why are you doing this, why don't you paint like you painted in the sixties or the forties?" And you have to say, "Yes, but I'm not the same person that I was in the forties. I have gone out into the world, and I've learned a lot, and I must express some of that."

Although Wilde would probably have denied this central aspect of reflectivity in art, like the aging portrait of Dorian Gray that reflected the sin Gray had committed, Lanyon's art reflected her maturation and experience.

Perhaps Wilde's "art for art's sake" approach to literature was a product of his time; he viewed his work as fine art, appreciated for its aesthetics without moral ramifications. Others have argued that Wilde was not completely dismissive of the moral repercussions of his work. In contrast, our interviewees looked at art as functional, especially for social, political, and educational causes. The artists connected their professional ethics directly with universal morality, slighting conventional morality. They saw their abilities as gifts that only they could contribute to the world. They sought to have an impact on the world around them by sending messages of hope, equality, and change through their art.

Conclusion

In reflecting on the moral dilemmas these creative individuals faced, we might conclude that their attitude toward morality also reflected the creativity with which they approached their work. Whether they were doing creative science or creative art, they were also creative in the ways they approached issues of good and evil. The model we developed in Table 4.1 to describe different forms of moral action suggests that most of these individuals started their careers with a moral attitude that might be characterized as ignoring conventional morality, ignoring universal morality, but strongly endorsing professional ethics. As they grew to maturity in their careers, many of them adopted a moral attitude that could be described as endorsing professional ethics and universal morality, while still being neutral toward conventional morality. In some cases, however, the attitude toward conventional morality might become dismissive; or in certain cases conventional morality might also be embraced.

We found no trace of the amoral creative individual in our in-depth analysis. At worst, our interviewees were guilty of time-hoarding: they might become obsessed with their professions, and thus selfish and protective of their time. But no matter how busy they were, broader concerns were never far from their minds. Almost all of the interviewees at some point endorsed both professional ethics and universal morality because these tend to grow from and give meaning to involvement in their professions.

Scientists and artists value their professional ethics highly, and many, though not all, reach out toward universality morality. The high regard for professional ethics is to be expected among highly creative individuals because their success rests on establishing and maintaining their reputation in their domains. The one form of morality they seem to abide by least is the conventional one. They ignore conventional morality if it conflicts with professional ethics or with universal morality, and have a propensity to defy or deride it. Conventional morality is usually the most volatile type and the one least firmly based on enduring values. Perhaps these creative individuals are aware of its often arbitrary nature, or maybe their values are simply ahead of their time.

This chapter has provided some evidence that creative individuals are morally grounded. They exhibit creative morality to maximize benefits and minimize harm in the consequences of their creative work. Like most stereotypes in the popular media, the amoral creative individual is perhaps just that, a misconception.

References

Bethe, H. A. (1991). *The Road from Los Alamos*. New York: Touchstone.
Choe, I. S., Nakamura, J., & Csikszentmihalyi, M. (1996). *The values of creative individuals*. unpublished manuscript.
Csikszentmihalyi, M. (1996). *Creativity: Flow and the Psychology of Discovery and Invention*. New York: HarperPerennial.
David, S. (writer), & Singer, B. (director). (2004). Occam's Razor [Television series episode]. In P. Attanasio, K. Jacobs, S. David, S. Bryan (producers), *House M.D.* Century City, CA: 20th Century Fox Studios.
Deigh, J. (2010). *Introduction to Ethics*. Cambridge: Cambridge University Press.
Marangoz, Ç. S. (2006). The idea of "delight and instruct" in Aristotle's poetics, Horace's ars poetica and William Wordsworth's preface to lyrical ballads. *Uluslararası Yönetim İktisat ve İşletme Dergisi, 2*, 211–219.
Martin, D. (2007, September 8). Madeleine L'Engle, writer of children's classics, is dead at 88. *The New York Times*. Retrieved from http://www.nytimes.com/2007/09/08/books/07cnd-lengle.html?pagewanted=all&_r=1&
Ogilvy, J. (1976). Art and ethics. *Journal of Value Inquiry, 10*(1), 1–6.
Rosalyn Yalow–Banquet speech. (1977, December 10). Retrieved from http://www.nobelprize.org/nobel_prizes/medicine/laureates/1977/yalow-speech.html

Sample, I. (2009, March 03). What is this thing we call science? Here's one definition. *The Guardian*. Retrieved from http://www.guardian.co.uk/science/blog/2009/mar/03/science-definition-council-francis-bacon

Schweber, S. S. (2000). *In the Shadow of the Bomb: Oppenheimer, Bethe, and the Moral Responsibility of the Scientists*. Princeton, NJ: Princeton University Press.

Shiner, L. (2001). *The Invention of Art: A Cultural History*. Chicago, IL: University of Chicago Press.

Stanford, S. (interviewer) & Gordimer, N. (interviewee). (2005). *Interview with Nadine Gordimer [Interview video]*. Retrieved from http://www.nobelprize.org/mediaplayer/index.php?id=418

Wilde, O. (1891). *The Picture of Dorian Gray*. London: Simpkin, Marshall, Hamilton, Kent. Retrieved from http://www.gutenberg.org/ebooks/26740, accessed November 15, 2013.

Wilson, E. (2006). *Naturalist*. New York: Warner Books.

5
Creative Artists and Creative Scientists: Where Does the Buck Stop?

James Noonan and Howard Gardner
Harvard Graduate School of Education, USA

Extending the systems view

Imagine this scenario: Picasso, an innovator of early twentieth-century modern art, is sitting in his studio in France in the 1950s. He has just received unsettling news from Spain. General Franco, the autocratic dictator long despised by Picasso, has given a nationally broadcast speech in which he praised the artist effusively. This bulletin alone, while distasteful, is not what is troubling him. Instead, he is replaying a passage from the speech in which Franco referred to one of Picasso's recent paintings and gave his considered interpretation. Believing—not entirely incorrectly—that Picasso had a longing to reconnect with the country he had not seen in decades, Franco suggested that this latest work reflected an important turning point in Picasso's career: unlike earlier paintings, this one represented deeply nationalist sentiments that had long lain dormant. Furthermore, Franco expressed his gratitude for Picasso's tacit but unambiguous endorsement of the regime.

For Picasso, this interpretation represented not only a complete distortion of his intent, but more dangerously, it risked co-opting his legacy and corrupting his reputation. That said, it was also true that Picasso *had* longed for attention and critical acclaim in his home country—so much so that he had briefly entertained accepting Franco's invitation to stage a retrospective exhibition in Spain. Nevertheless, Picasso knew that he had to speak out—to set the record straight and put distance between him and Franco—but he also needed to do so in a way that could preserve his artistic reputation in Spain and elsewhere.

This scenario represents one of many possible ethical dilemmas facing creative individuals: what to do when their work has not only been misunderstood but also misused and even abused. Across disparate domains, creative individuals tend to feel driven by a sense of responsibility to their ideas rather than to what others in the craft may have done. This devotion to ideas—and in particular, ideas that diverge from conventional thinking and ways of working—has at times transformed domains. Indeed, Picasso's impact on the visual arts was enormous: it changed the way painters, sculptors, and other visual artists approached their work and helped to define the modern era (Gardner, 2011a). More recently, one need only look to the multitudes of creative individuals populating the emerging high-tech industry for contemporary examples of domain-creating and domain-changing innovation.

These examples reflect a systems view of creativity. Instead of focusing on creativity simply or primarily as an individual characteristic, this perspective sees it as a dynamic process entailing the domain and the context in which individuals work (Csikszentmihalyi, 1996; Feldman, Csikszentmihalyi, & Gardner, 1994). In its most specified form, Csikszentmihalyi (1996) proposed that creativity—and the ideas and products born from creativity—emerge from the continual interaction of three nodes: creative individuals, the domain in which they work, and the reactions and actions of peers working in the same sphere. In most cases, once the creative contribution has been put into the world and validated by the field, the creative individual does nothing more. The creative ideas continue to evolve, occasionally inspiring subsequent innovations, but by that time the creators have often moved on to their next project.

Despite the often frenetic pace of innovation, we assert that creative individuals cannot, or at least should not, completely abdicate responsibility for their contributions. As such, we extend the systems view of creativity to consider what happens when a creative individual (or his or her contribution) is affected by subsequent events—what we are calling a post-creative development (hereafter PCD). This development could be a threat, as in the imagined scenario above, and it could take multiple forms: another individual, a group, a prevailing ideology, or a social movement. In order to be threatening, the PCD must be sufficiently *powerful* (politically, financially, or otherwise) to influence public perceptions or attitudes toward the contributor or to the creative achievement in such a way that the contribution could be misunderstood, misused, or abused by others. For example, the imagined threat from Franco took the form of both a powerful individual and a comprehensive ideology. His influence—in Spain, at least—was nearly omnipresent; accordingly, he would have been able to make outrageous claims and imbue them with some credibility simply by virtue of speaking them out loud. However, a PCD could also be an opportunity. For example, a television network or an automobile company could decide to use Picasso's Cubist vision as the basis for its public brand, thereby bringing his work to the attention of millions of people who might otherwise not have encountered it.

In this chapter, we extend one of the nodes from Csikszentmihalyi's (1996) schema and consider it through the lens of ethics. In focusing on ethics, we make a deliberate contrast to mere politeness and morality, which define social expectations and guidelines for proper behavior. We also intend a contrast to obedience to the law, which sets a minimum expectation for all individuals in a society. We reserve the term and the realm of *ethics* for cases in which specific responsibilities are associated with specific roles. The prototypical cases here are the learned professions. We do not expect doctors or auditors or lawyers simply to behave cordially and to obey the law. Rather, over the years, we have come to associate and expect certain attitudes and behaviors on the part of those professionals—for example, physicians should not recommend treatment in which they have a financial investment, auditors should not have personal relations with their clients, lawyers should consider themselves as officers of the court. Applying this conception to the realms under consideration, we explore here whether there is a unique responsibility for individuals occupying the role of creator, regardless of the profession or line of work that they have chosen.

Accordingly, the dilemmas we discuss in this chapter pertain to questions of the right thing to do in the context of individuals' work. More specifically, we examine the relationship between creative individuals, their creative contribution, and the PCD facing them. Although there is a temporal aspect to this relationship—with the individual preceding the contribution, and the contribution preceding the PCD—we contend that both the individual and the PCD have unique relationships to the creative contribution. For this reason, as noted in Figure 5.1, we are focusing on the association between the creative individual and the development that follows his or her creation. Because creative individuals operate across diverse domains, we illustrate these associations using examples from the arts and sciences. These illustrations help us to demonstrate our central contention: when creative contributions across domains face developments that could either threaten or benefit them, it is the ultimate responsibility of the creative individual (1) to

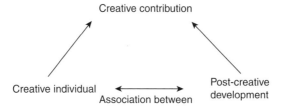

Figure 5.1 Extending the systems view of creativity to consider post-creative developments (PCDs).

identify the development (as either a threat or an opportunity); (2) to assess options for how to respond; and (3) if necessary, to take steps to counteract it. In what follows, we address both types of developments. We attend more closely to the threats since they place creators in a more vulnerable position where their ability (or inability) to respond is especially instructive. Further, we contend that appropriate responses are easier to effect when creators are members of an established profession, but that in no case is the creator relieved of their responsibility to respond.

The dimensions of ethical creativity

While there are many valid and instructive ways to analyze the ethics of creativity, we have chosen to focus on the two dimensions that demonstrate essential contrasts: (1) whether creative individuals *benefit* from the association or whether they are the *victim*; and (2) whether these creative individuals are *active* or *passive* participants. Let us illustrate these contrasts initially with the help of another scenario, this time with an imaginary protagonist as well as an imaginary scenario. Rosalind is a research scientist interested in genetics, and she has recently made a breakthrough in her work on DNA sequencing. First, it is important to recognize that creative individuals like this scientist may sometimes derive *benefits* from even the most ethically murky of associations. These benefits may be conferred on them in the form of financial windfalls or greater visibility for their work. Let us imagine that Rosalind learns that the military is interested in her work and has offered her a position as a high-level advisor. This position could afford her considerable influence in and beyond her field, not to mention potential financial benefits. Of course, it is also possible that she could be *victimized* by this association. If, for example, it turned out that the military wanted to adapt her innovation for use in developing biological weapons—or even if the public or the scientists' colleagues *believed* this to be the case—then this association could injure her reputation and career.

Most ethical dilemmas facing creative people lie somewhere between the two extremes of this dimension. Potential benefits are often counterbalanced by potential costs. Similarly, an association that may seem initially beneficial (an opportunity) may wind up being enormously destructive (a threat). For example, consider what happens when creative individuals associate themselves (or become otherwise associated) with political regimes and then the regime or administration changes. (Other changes in political contexts can also prompt creative individuals to abandon once promising associations. For example, the Arte Povera movement in Italy in the late 1960s initially attracted numerous adherents with the notion that art should be created from common materials. Once the movement became associated with increasingly radical and violent student demonstrations, however, the association no longer seemed productive. Many artists, including most prominently Alighiero e Boetti, consciously sought to distance themselves from the movement; indeed, Boetti even changed his name. For him, nevertheless, this distancing resulted in a very successful reinvention, inasmuch as he is now best known for the work he created *after* the Arte Povera movement (see Cotter, 2012).)

Regarding the second dimension, let us return to the imaginary scenario featuring Rosalind, the geneticist. Where we left her, she was facing an *active* choice about whether or not to take the offer as a military advisor; that is, she had some agency over how to manage her association with the military. She needed to determine whether this association was potentially threatening and, if so, whether the potential benefits outweighed any potential risks. This type of decision-making is, in many ways, a luxury. One could also imagine a scenario in which Rosalind published an influential book about her findings and might simply have her book bought en masse by the military with the intent of mining its contents for potentially destructive adaptations. In this case financial benefits would accrue to her *passively* through book royalties and her sphere of influence would be much more limited. After all, she would have no direct audience with the consumers of her ideas. However, it would be misleading to say that Rosalind would have no influence at all. As the originator of a creative contribution, she would retain considerable influence by virtue of her "bully pulpit." In the event that an innovator becomes aware that an innovation may be not only misunderstood but misused or abused, she or he faces an ethical choice about whether, when, and how to make use of this influence.

Table 5.1 Creative individuals facing post-creative developments along two dimensions

	Active	Passive
Beneficiary	I The Opportunist	II The Reluctant Winner
Victim	III The Unlucky Gambler	IV The Hostage

As already noted, the association between a creative individual and a threat may take many forms. In Table 5.1, we present four scenarios in which creative individuals might find themselves. In Quadrant I, we find the *Opportunist*, an active beneficiary who willingly associates himself or his work with a PCD in exchange for what he assumes will be positive consequences. In contrast, in Quadrant II, the *Reluctant Winner* recognizes that while an association with a PCD may extend some benefits to her, she is either unwilling to associate herself actively or she has no control over the association. In Quadrant III, the *Unlucky Gambler* (who may have started out as an Opportunist in Quadrant I) willingly associates himself or his work with a PCD, but this association turns out to have negative consequences. In Quadrant IV, the *Hostage* has her work unwillingly associated with a PCD, either as a paragon (as in our imaginary example with Picasso and Franco) or possibly as a foil to be demonized—and this association has negative consequences.

In subsequent sections of this chapter, we first provide idealized portraits of a creative individual who might typify each quadrant. Moving away from "ideal types" (Weber, 1958), we supplement these portraits with snapshots of artists or scientists who faced similar dilemmas in their work.

Quadrant I: The Opportunist

Mr. A is a painter living in a country that has recently experienced considerable social upheaval. Inspired by the sweeping rhetoric of a candidate for national office, Mr. A begins experimenting with new media. Without his having expected it, his dynamic works, filled with popular images, quickly become picked up and peddled by the candidate's campaign. As a result, Mr. A's profile as an artist is elevated considerably. In addition, he is given new commissions and high-visibility art shows at nationally respected museums. He does nothing to distance himself from the campaign; on the contrary, he promotes it. Some call his work "propaganda," but Mr. A says that his work is merely the result of his personal inspiration. He willingly decides to produce new work that the campaign can use in fundraising. By first allowing his work to be associated with a PCD (in this case, a campaign that sees its message reflected in the art) and then actively pursuing this association, Mr. A's financial prospects and visibility benefit.

Like many other people, graphic designer and illustrator Shepard Fairey first heard Barack Obama speak at the 2004 Democratic National Convention and liked what he heard. In 2008, with the Democratic primaries heating up and early indications that Obama and Hillary Clinton would be engaged in a prolonged fight for their party's nomination, Fairey ran into Los Angeles publicist Yosi Sergant at a party and shared his support for Obama. At the time, Sergant was working as a media consultant for the Obama campaign; he suggested to Fairey that he do something as an artist to show his support. The next day, Fairey called him to ask if the Obama campaign would mind if he made a campaign poster (Mcdonald, 2008). Since he had been arrested 14 times related to his guerrilla-style street art and was aware that he could be a potential liability to the campaign, Fairey wanted to get the go-ahead from someone close to the campaign. In addition, much of his recent work had been critical of capitalism and nationalism, and he conceded that "a person who is a blindly nationalistic type could try to spin my work as being un-American or unpatriotic, and I was afraid that … my poster for Obama could be perceived as the unwelcome endorsement" (Shorrock, 2009).

Nonetheless, two weeks before Super Tuesday, the campaign gave Sergant (and Fairey) its tacit endorsement. In less than a week, Fairey designed and printed the first run of his now-iconic HOPE poster, distributing 350 on the street and selling another 350 through his website. (The initial run of posters actually featured the word PROGRESS as a caption. In an interview after the election, Fairey explained that he received feedback from people that "hope was the message that the Obama campaign really wanted to push" (Shorrock, 2009).) The image-reproduced on posters, stickers, and T-shirts and shared through social media outlets—soon went viral, spawning numerous caricatures and imitators. The visibility and positive press immediately injected Fairey's name and reputation into the mainstream. Within months, buoyed by an overwhelming public response to the poster, the campaign commissioned "official" posters, done in the same style but using photographs that it had clearance to reproduce; moreover, campaign managers began selling the posters through the official campaign website for $70 each. Fairey also received a personal letter from Obama thanking him for "using his talent in support of my campaign" (Shorrock, 2009). The benefits to Fairey of his association with the campaign were considerable. While many of his posters were reproduced and distributed at no cost, he sold many more through his website (although he contended that any proceeds were reinvested in producing more posters or donated directly to the campaign). Perhaps more enduringly, though, Fairey earned name recognition from the poster that he assuredly would not have had otherwise. He also earned credibility that transcended his reputation as a street artist or graphic artist, with *Time* magazine commissioning a version of the image for its Person of the Year cover at the end of 2008 and the campaign commissioning an original design for its official inauguration poster. In addition, in early 2009 an original print of the HOPE image was obtained by the Smithsonian Institution's National Portrait Gallery; soon thereafter, Fairey had his first career retrospective solo art show at the Institute of Contemporary Art in Boston (Edgers, 2009).

By inserting himself so explicitly into a national campaign and its accompanying spotlight—and as a result of the attention he garnered—Fairey also invited increased scrutiny. This scrutiny included a widely publicized legal dispute with Associated Press over copyright infringement that was initially settled out of court, but that reemerged after it was discovered that Fairey had destroyed evidence in the case (Kennedy, 2012). Even so, Fairey sought to use the negative press to push a conversation about fair use and "artistic freedom"; his views on the topic surely attract more attention (and arguably his opinions carry more weight) as a result of his well-established association with Obama (Fairey, 2012).

Although the artistic style embodied by the HOPE poster was not new to Fairey, associating himself and his work with a highly visible public figure represented a clear departure from his previous work and way of working (which was characterized by posters or stickers clandestinely—and often illegally—installed in public spaces). This shift, and Fairey's association with the PCD of the Obama campaign, conferred on him considerable benefits. Indeed, the iconic image ensured future commissions and future potential consumers. Even the legal actions against him seemed unlikely to blunt these benefits.

Quadrant II: The Reluctant Winner

The B's are a popular band whose new genre-bending song has received considerable airplay and elevated the band's public profile. Seeking to capitalize on The B's popularity with the youth demographic, the widely popular President mentions the band and their song by name in a nationally televised speech that is then rebroadcast on several networks. The unspoken implication in the President's remarks is that The B's support his policies. Although personally opposed to the President and his policies, The B's recognize that this association may have tremendous financial benefits for them. It could also expand their fan base. In addition, taking a political stand against the popular President could very well alienate legions of existing fans. After some deliberation, they opt to adopt a "wait and see" approach and do nothing. Although they did not seek out the benefits they are likely to receive, they decide that an association with a popular public figure is a post-creative development worth accepting.

On more than one occasion, the country songwriter and recording artist Gretchen Peters found herself a Reluctant Winner; in both cases she found ways to make peace with the benefits she received. In 1994, Peters wrote a song called "Independence Day" about a woman in an abusive relationship and was entitled to her share of the proceeds any time the song received airtime or was used in advertisements. Peters also was transparent about her politics, aligning herself comfortably with left-wing or progressive causes. A dilemma arose when conservative commentator Sean Hannity began using "Independence Day" as the lead-in song on his popular nationally syndicated radio show.

Peters had a decision to make. On the one hand, Hannity's use of her song meant that she received considerable financial royalties. In addition, the national exposure she received could indirectly lead to an increase in record sales from Hannity listeners who liked her music. On the other hand, Peters recognized that Hannity had either misinterpreted or willfully ignored the intent behind the song; a song that was not, as he seemed to think, a paean to American liberty, but in fact a rallying call for abused women. Furthermore—and perhaps more importantly—the association between Peters' song and Hannity's show, if left uncommented on, could be interpreted as an endorsement by Peters of Hannity's politics. In this way, Peters recognized the potential of the association with Hannity as a potential threat to her reputation. She also understood that it was an association that she did not seek, but that could—depending on the decisions she made—benefit her. With these considerations in mind, Peters decided to do two things. First, she decided—unlike many recording artists whose songs are employed for political candidates or causes they find disagreeable[1]—*not* to issue any cease-and-desist orders. On the contrary, she encouraged Hannity's use of the song for his radio show. Second, she announced publicly that any proceeds she received from Hannity's use of the song would be redirected into political causes of her choice. (These included the American Civil Liberties Union and the political action website MoveOn.org.) By making a public statement, Peters asserted her influence over her work as the creative individual. She employed similar tactics during the 2008 presidential race when the song was used at a rally for vice presidential candidate Governor Sarah Palin. In her public statement about the song's use, Peters said the campaign was "co-opting the song, completely overlooking the context and message and using it to promote a candidate who would set women's rights back decades" (as quoted in Shelburne, 2008). As with her response to Hannity, Peters announced that she would donate the royalties received during the election cycle to Planned Parenthood, and do so in Sarah Palin's name. Through these choices, she was able to accept the windfall that her song's association with PCDs bestowed on her, while also preserving her sense of artistic and creative integrity.

We recognize that the decisions Peters made could be perceived as an "active" stance, but we note that these decisions reflect her *response* to the PCD. We see Peters as a passive beneficiary, because she did not seek out any of the benefits accruing to her. Once she had realized that her work was associated with a PCD, though, she had decisions to make. We contend that all creative individuals—whether they actively seek out a relationship to a PCD or whether the PCD finds them instead—must undergo a similar reflective process about whether and how to respond.

Quadrant III: The Unlucky Gambler

Ms. C is a popular and well-respected filmmaker. She has made feature films and documentaries that have received widespread critical acclaim for their innovative techniques. Among Ms. C's very public fans is the increasingly autocratic leader of her country. Initially flattered and grateful for the exposure that this association brought her, Ms. C sought

> *out the attention of the leader and arranged private screenings for him. However, recently she has become alarmed by statements that she considers aimed squarely at reducing her artistic independence. For example, in a recent speech the leader referred to works as "my movies." Furthermore, he dismissed her earlier body of work—films that had earned accolades from around the world and sealed her reputation as an innovative filmmaker—as "experimental garbage." The association is not only no longer fruitful, but also it seems that it may threaten her career and livelihood.*

The classic—some might say clichéd—example of an Unlucky Gambler is the German filmmaker Leni Riefenstahl. Originally an actress, Riefenstahl had her directorial debut in the 1932 film *The Blue Light*, a work that received critical acclaim and played to full houses across Europe. However, she is best (and most notoriously) known for her association with Hitler and the Nazi Party in the years before the Second World War. According to Trimborn (2007), Riefenstahl sought out the association. After being mesmerized by a speech at a Nazi rally in 1932, Riefenstahl wrote a letter to Hitler and requested a face-to-face meeting. To her surprise, and delight, he accepted her request and the two met. During this meeting, Hitler allegedly implored her that, should the Nazis come to power, she "must make my films" (as quoted in Trimborn, 2007, p. 60). When Hitler became Chancellor the following year, he enlisted Riefenstahl to film Nazi rallies in Nuremberg in 1933 and 1934; the latter, released as a film called *The Triumph of Will*, became widely considered a model of Nazi propaganda.

However, Riefenstahl would perhaps be far less renowned if she was remembered solely as a propagandist. In addition, she was widely credited as an innovative and path-breaking documentary filmmaker; the primary case in point being her two-part documentary of the 1936 Berlin Olympics. Buoyed by government funding and support at the highest levels of the Nazi party, Riefenstahl spared no expense, traveling to Greece to film the torch-lighting and relay, using stop-gap techniques to have actors appear to morph from ancient Greek to contemporary athletes, putting cameras on cranes or tracks to better document motion, and employing slow-motion photography (Trimborn, 2007). In many ways, she was a predecessor of contemporary sports photography. Riefenstahl's active pursuit of an association with Hitler and the Nazi regime (in this case, the PCD) led to nearly unlimited resources and unparalleled distribution, both of which enabled her to experiment with techniques that burnished her reputation as an innovator. When the war ended, though, so too did the benefits this association extended to Riefenstahl. When she was detained and interrogated by American soldiers, she struggled to reconcile the horrors of the Holocaust with "Hitler, as I knew him" (as quoted in Bach, 2008, p. 226). Indeed, she insisted throughout the rest of her life that the association was innocuous and that she was uninterested in politics. In this way, Riefenstahl is a prime example of a creative individual who could not capably respond to PCDs, because she was never able (or never willing) to identify them as threatening. By distorting the influence of Hitler and Nazism on her work, she undermined her own reputation. For that matter, this refusal to acknowledge the threat posed by Hitler was responsible for the end of her film career. Lacking the support and patronage of the Nazi party, it took Riefenstahl another 10 years to complete the film she was working on when the war ended. However, many film festivals refused to show it and she never made another, turning instead for her livelihood to photography and memoir writing.

Another example: Many of the scientists working on the Manhattan Project in the early 1940s—like Eugene Rabinowitch, Joseph Rotblat, and Hans Bethe—believed that associating themselves and their expertise with the US government was not only necessary but also the right thing to do under the circumstances. Fearful that Germany might unlock the potential of atomic energy and develop and use nuclear weapons, these scientists were convinced that it was imperative to win the arms race. In this case, the PCD might be characterized as the perceived German threat and the US military's perceived need to counteract it, which came to be known as the Manhattan Project. The association of scientists' expertise with the military motives of the US government was enormously productive, although it was hardly beneficial to the scientists in conventional ways. They received no financial windfall for their participation. Moreover, since the project was top secret, they received no boost to their reputation either. Their work was seen as an act of unselfish patriotism.

However, as the war proceeded and it became apparent that Germany was not working to complete an atomic bomb, many scientists began to question this association. Rabinowitch and a vocal cohort of colleagues on the Manhattan Project clearly recognized the potential negative side effects of the association; accordingly, they formed a secret committee of scientists associated with the project. The so-called Interim Committee met in secret for a week in early June 1945 to draft what came to be known as the Franck Report, named after Nobel Prize winner James Franck (Price, 1995). Rabinowitch took primary responsibility for drafting this widely heralded treatise, which recommended that atomic weapons be put in control of international bodies and never used on civilian populations. Delivered to the Secretary of War in June 1945, the Franck Report urged that nuclear weapons be demonstrated "before the eyes of representatives of all the United Nations, on the desert or a barren island" and urged the United States to refrain from using the atomic bomb on Japan, calling an early and unannounced attack "inadvisable" (Dannen, 1998).

While having little if any impact on government decision-making, the Franck report foreshadowed an increased political awakening for many scientists. In 1945, Rabinowitch co-founded the Atomic Scientists of Chicago (later the Educational Foundation for Nuclear Science), an organization open to "[a]ny past or present member of the Manhattan Project" and whose goals were to "clarify... the responsibilities of scientists in regard to the problems brought about by the release of nuclear energy" and to educate the public on these matters (Atomic Scientists of Chicago, 1945). (Perhaps the best-known initiative of the *Bulletin* is the Doomsday Clock, a meme meant to convey "how close humanity is to catastrophic destruction"; see http://www.thebulletin.org/content/doomsday-clock/overview). In 1952, Rabinowitch wrote that "the explosion of the first atomic bomb... led to a decisive change in the political consciousness of scientists" (Rabinowitch, 1952, p. 314). Scientists like Rabinowitch "gambled" on a PCD in that they willingly associated their expertise with seemingly noble and patriotic aims. The association brought them no benefits beyond intrinsic rewards, and the signers of the Franck report worried that the association could imperil their reputation. To counteract this perceived threat, Rabinowitch and others drew on their sense of professional ethics (Slaney, 2012), a notion to which we return later in this chapter.

Quadrant IV: The Hostage

The recently published book of poetry by Ms. D was an international sensation. In its treatment of global climate change, the verse was both innovative and evocative. It spawned several imitators, and initially seemed as if it were changing the way in which poets and artists of all stripes portrayed the environment. Within the year, however, corporate lobbyists in her country undertook a campaign to discredit her, calling

> her a propagandist on behalf of environmental groups and describing her
> poetry as anathema to national values. *In many states, similar localized
> campaigns resulted in her book being publicly pilloried and even banned
> from schools. Before long, other artists began to distance themselves
> from Ms. D. Her prestige and financial security evaporated. Reluctantly,
> so as to distance herself from the excoriation, she emigrated. The new
> start, she hoped, would enable her to continue writing and publishing.
> Ms. D did not seek to be a spokesperson for any social movement; rather,
> this perception emerged out of an association between her work and a
> PCD. The backlash against this perceived association—unsought and,
> in Ms. D's case, undesired—wound up imperiling her reputation and her
> livelihood.*

Erich Maria Remarque was conscripted into the Germany army in
November 1916. While little is known about his experiences as a soldier,
his war novel, *All Quiet on the Western Front*, quickly became an international sensation when it was published in 1929. Widely considered
one of the finest novels about the First World War, the book was initially refused by several publishers who believed that readers were not
yet ready to relive the war years. This skittishness on the part of publishers may have been in part because Remarque's view of the war was
decidedly less than glorious, presenting a vivid and disturbing picture
of a "lost generation" of young people disillusioned and psychologically
scarred by their war experiences. However, contrary to the publishers'
expectations, this picture resonated with thousands of returning soldiers
and with a broad readership of war-weary individuals on all sides of the
conflict. In his first-person novel, Remarque presented readers with a
tangible manifestation of the otherwise "unknown soldier" who acted
as an Everyman, enabling readers to imagine that their suffering (or
that of their families or friends) was shared. Cultural historian Modris
Eksteins remarked that "[t]he great discovery that foreign readers said
they made...was that the German solder's experience of the war had
been, in its essentials, no different from that of solders in other nations'
(1990, p. 296). To say the least, it was a narrative that did not lend itself
well to the patriotic stirrings of a surging tide of nationalism.

The book's meteoric rise was soon matched by an equally swift fall,
at least in Germany. It was quickly taken up as a foil by the ascen-
dant Nazi Party, which prioritized rearmament and relied for its support
on a renewed and virulent strain of nationalism. The German military

referred to the book as pacifist propaganda that was "a singularly monstrous slander of the German army" (quoted in Eksteins, 1990, p. 288). In 1930, one Nazi state minister banned the book from schools; and after Nazi hooligans disrupted showings of the film adaptation, it too was banned, ostensibly for the protection of the German people. Once the Nazis came to power in 1933, Remarque's books were burned as "politically and morally un-German" (quoted in Eksteins, 1990, p. 298). By the end of that year, copies of the book were seized from the publisher and ordered to be destroyed by Minister of Propaganda Joseph Goebbels.

In such a volatile political climate, creative individuals confront the threat of having their work unduly influenced or suppressed altogether. Like many of his contemporaries, Remarque faced a choice: given this threat, how could he respond in a way that preserved his reputation and that of his work? In 1930, a little more than a year after the book's publication and with signs that its association with the Nazi party was a potentially negative PCD, Remarque sought refuge in Switzerland (with detours to New York and Hollywood, where he helped adapt his book for film). Remarque's decision to seek exile abroad was based in part on the decisions made by many creative individuals living under autocratic regimes: only by leaving the context in which his creative contribution (and, by extension, his reputation as a writer) was being abused could he effectively respond to the threat that this association posed. The association between the Nazi party (in this case, a PCD) and himself was neither sought nor beneficial to Remarque; only by putting physical distance between himself and the Third Reich could he blunt the threat. As a result of this distance, he was able to continue writing, but, just as importantly, he was able to preserve his and his book's reputation. (An even more esteemed writer, Thomas Mann, made similar career and life decisions.)

An example from a quite different realm: After years of reanalyzing data culled from a national survey on social capital to try to find alternative explanations, political scientist Robert Putnam shared what he learned: communities with greater diversity tend to have less trust among neighbors, lower rates of voting, lower likelihood of donating to charity (see Putnam, 2007). Indeed, a spectrum of indicators of civic health appeared to be lower in more diverse communities. This finding seemed at odds with Putnam's reputation as a liberal academic that placed him "squarely in the pro-diversity camp" (Jonas, 2007). Not surprisingly, these dystopic results quickly attracted attention from conservative politicians and anti-immigration groups, including a favorable commentary on the website of former Ku Klux Klan leader David Duke, as proof that large-scale immigration had a deleterious effect on the fabric of American life. This co-optation of his findings posed a threat to the substance and integrity of Putnam's work (and, conceivably, to his reputation as a social scientist).

Anticipating that his findings would be co-opted in precisely this way, Putnam did two things. First, he supplemented his findings with commentary about how communities could, with time and judicious policy intervention, overcome the negative correlation between diversity and social capital. He suggested that while "in the short run there is a tradeoff between diversity and community...wise policies (public and private) can ameliorate that tradeoff" (Putnam, 2007, p. 164). In fact, in an interview with the *Financial Times*, Putnam asserted that he delayed publishing the research until he could develop proposals to counteract the negative effects of diversity, saying that it would have been "irresponsible" to publish his article without such commentary (Lloyd, 2006). This action was seen as suspect by some who claimed that the role of social science researcher required one to maintain a disinterested stance and refrain from social commentary. However, others viewed Putnam's publication (even with the coda) as an act of professional responsibility: that he had an obligation to share inconvenient truths and did so. Second, once the commentaries from people like Duke came as predicted, Putnam tried to respond in kind, giving interviews and making public appearances to press his point. Rather than leave his work to the world of commentators—both those who understood his intent and those who abused it—Putnam made the decision to be an active participant in the debates and discussions prompted by his findings. Having not sought out the association with a PCD (in this case, widely derided commentators like Duke), Putnam nevertheless recognized that without intervention this association posed a threat to his reputation as a social scientist.

In this way, Putnam's actions were compatible with the actions taken by two previous examples. Gretchen Peters (in our parlance, a Reluctant Winner) similarly perceived that an unsought association with a PCD—even though it conferred on her financial and other benefits—demanded intervention. Similarly, Eugene Rabinowitch (an Unlucky Gambler) determined that his active association with the Manhattan Project required recasting if he was to preserve his reputation and set a model as an ethical scientist.

Crossing quadrants: Evolving PCDs and evolving responses

As we noted earlier, the neat classification of creative contributors into four quadrants is useful for understanding the phenomenon that we have been investigating. However, the formulation does not capture the fact that creators' association with PCDs, and their attendant responses to them, often evolve over time and across contexts. Several of the examples above could also be used to illustrate this point, but as it happens, one of the authors of this chapter (Howard Gardner) has had the opportunity to observe post-creative developments in his own professional life.

As is well known, at the beginning of the twentieth century the French psychologist Alfred Binet devised the first intelligence tests. His goal was clear: to allow the early detection of potential scholastic problems and to encourage interventions that would help an intellectually challenged population of young people to succeed in school. Shortly after intelligence tests were devised, they became standardized (and popular) in the United States (and other countries). Rather than being used as an early warning sign, however, Binet's tests (and their offspring) quickly came to be used as ways of classifying students altogether. Those who did well were treated as academically talented and were often given special opportunities, such as enrolment in programs for "gifted and talented youth." Conversely, those who performed poorly were often considered uneducable: they frequently received less skilled teaching and sometimes they were segregated from peers and deprived of teaching altogether.

Many years later, Gardner was part of a group of psychologists who were critical of standard intelligence theory and regular intelligence tests. In Gardner's case, as part of his theory of multiple intelligences (MI theory), he proposed that human cognition is better described as a set of relatively autonomous intellectual faculties, which he dubbed the multiple intelligences (Gardner, 1983). He also resisted the urge to declare these faculties as immutable; indeed, on the basis of scattered evidence, he argued that any intellectual faculty could be enhanced through a combination of good teaching, adequate pedagogical resources, and strong motivation on the part of the learner.

While receiving considerable criticism from within the psychology (and particularly the psychometric) community, MI ideas were quickly embraced by educators; initially in the USA, ultimately in many other countries in the world. At the beginning, Gardner was the *beneficiary* of these post-creative developments; they added both to his fame and (modestly) to his fortune. Gardner was mostly a passive beneficiary (Quadrant II): he did not proselytize, but he sometimes benefited even when he was not in sympathy with his benefactors.

However, a wake-up call occurred in the early 1990s. Gardner learned that, in an Australian state, young people were being classified in terms of the intelligences that they possessed and ones that they lacked. This claim went well beyond Gardner's own research. The wake-up call turned into a smoking gun when the Australian project listed the major ethnic and racial groups in the state, along with a claim about which intelligences they allegedly possessed and which they allegedly lacked. Not only was there no empirical evidence for this claim, Gardner also saw the claim as potentially dangerous, because it could become a self-fulfilling prophecy and as a stimulus for destructive social policies. This event, along with a few other similar ones, caused Gardner to change his stance considerably. He could no longer be a passive beneficiary; the association between his ideas and how they were being used had changed significantly. Because his ideas were being misused and abused, the integrity of the ideas was at risk (and so, too, was Gardner's reputation). He was still passive in that he had not sought an association with the Australian state, but he was now at risk of being victimized by this association if he did nothing (Quadrant IV).

After reflecting on the PCD, Gardner responded in three ways. To begin with, he went on television in Australia and critiqued the educational intervention; soon thereafter it was cancelled. Then, too, he wrote an article in which he separated out the myths and realities of MI theory (Gardner, 1995); the article was reprinted in many venues, and remains the most often cited by Gardner's colleagues in education. Third, and most significantly, within a year or two, working with distinguished colleagues, Gardner launched a project that was initially called "humane creativity." A particular challenge for the project was to determine how one could help to ensure that creative inventions would be used responsibly. As part of this project, Gardner and colleagues now spend much of their time trying to help individuals direct their skills— creative and otherwise—to positive, beneficent use. In our terms, they have shifted from being reluctant beneficiaries to becoming active users of the bully pulpit. One of the key findings of this project, now called the GoodWork Project, now called the Good Project (see thegoodproject.org), has been to understand the contributions of professional status to ethical and responsible decision-making (Gardner, Csikszentmihalyi, & Damon, 2001); it may also have helped to catalyze this volume.

The mediating role of the professions

Creative individuals can be found in any domain: in the arts, in business, in commerce, or in a well-established profession. Being part of a profession bestows on its members considerable benefits: for example, the prestige that accompanies a trustworthy and widely known credential (like admission to the bar or a medical degree). However, these benefits are not without costs. When individuals are inducted into an established profession, part of the cost of admission is their implicit agreement with the ethical strictures that govern that profession. Ethical codes guide professionals in determining the "right thing to do" when their personal interests or values are in conflict with the professional standards. Gardner (2011b) calls these professionally derived strictures the *ethics of roles* and argues that they are of increasing relevance for professionals. To use one of the above examples, Putnam is a political scientist. As such, he is enjoined to carry out his research objectively and to report his findings in a straightforward way. And once his "creative work" has gone public, he is expected to share his data, to explain his findings, and, as warranted, to incorporate new data into his conclusions.

In contrast, artists do not enjoy the benefits (nor the requirements) of the professions. Despite the advent of numerous credentialing mechanisms—for example, the proliferation of university-based MFA degree programs—these credentials are not required to be an artist. Anyone can make art. As such, artists, like other creative nonprofessionals, have no professional responsibilities beyond those of citizenship: minimally, they need to obey the law and try not to hurt anybody. Although it has a long history and many traditions, painting (like other forms of artistic production) is not a profession. Being a scientist used to have more in common with the artist. Tinkerers and amateurs like Michael Faraday and Alfred Russel Wallace did not belong to recognized professional cohorts, nor did they have any responsibility to their fields of study as such. (Despite having only a basic formal education, Faraday discovered electromagnetic induction, which in turn led to the widespread and practical use of electricity. Wallace, originally a surveyor, is credited as the co-discoverer of the theory of evolution.) These nineteenth-century "gentleman scholars" could make discoveries, put them into the world, and had little professional responsibility beyond that. But today, with myriad subspecialties and established credentialing agencies, more firmly established responsibility accompanies scientists' professional status.

The ethical standards that accompany professional status provide critical guideposts in helping individual workers to resolve ethical dilemmas. On the face of it, this state of affairs would suggest that creative contributors working in professions are better positioned to make ethical decisions than their counterparts in the creative non-professions, with professionalism acting as an important bulwark against ethical misconduct. However, this inference may go too far. As we have argued above, in responding to threats and resolving ethical dilemmas, creative individuals *across domains* should undertake a similar discernment process. At most, the clarity of roles and role requirements offer professionals an additional *resource* in resolving dilemmas. Creative non-professionals must rely more heavily on informal social networks and their individual discernment. Accordingly, there is likely to be greater variability in how non-professionals respond when confronted with similar ethical dilemmas.

In our view, irrespective of the domain in question, creative individuals must undertake a similar discernment process when their work is associated with a PCD; see Figure 5.2. First, they must *identify the development as such*, whether it be regarded as a threat or more as an opportunity. Ideas and products can be put into the world and used in innumerable ways. Often these ideas form the foundation for subsequent innovation, as Einstein's theories of relativity led to new ways of conceptualizing space exploration and our understanding of the origins of the universe. The vast majority of these uses are not threatening and so do not require a response from the creative individual. However, when innovations are susceptible to misinterpretation, misuse, or abuse, it is imperative that the association be first identified as such.

Such turns of events may not be immediately apparent, as in the case of Einstein's initial decision to catalyze the US government's efforts to build an atomic bomb. Approached by his colleague Leo Szilard who was concerned about unsecured uranium deposits in Congo, Einstein believed he had a responsibility to alert the government to

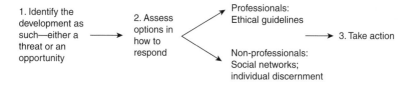

Figure 5.2 Process of responding to post-creative developments.

the possibility of Germany's using uranium to build an atomic bomb. Accordingly, in 1939, he sent a letter to President Roosevelt to this effect. Later, Einstein realized that his association with the Manhattan Project could pose a threat not only to his ideas and their legacy, but more importantly to those populations on whom the bomb could be used. He swiftly moved on to the next step for creative individuals responding to threats: he sought to *assess his options in how to respond.*

This assessment is where the difference between professionals and non-professionals is clearest. For individuals working in established professions, there are often ethical codes and role requirements that provide important guidance. For non-professionals, the lack of clarity leads to greater variability. In Einstein's case, by the time he was confronted with this dilemma, physics had become an established profession with numerous professional associations. In fact, Einstein even served as president of the German Physical Society—physics' oldest professional association—from 1916 to 1918. However, nuclear physics was an emerging subdomain within physics and the debate over how nuclear fission could or should be used (especially in the service of the war) was still an open question. (By the end of the war, with the benefit of hindsight, this was a more settled question. Many of the physicists and chemists who had participated in the Manhattan Project, like Eugene Rabinowitch, mentioned earlier in this chapter, drew on their sense of professional ethics to oppose the nuclear arms race.) Accordingly, in addition to his professional judgment, Einstein also looked to his social network to assess his options, writing letters of concern and seeking counsel from colleagues like Leó Szilárd and Niels Bohr (Isaacson, 2007).

Finally, having assessed his options, Einstein needed to decide just how to respond and then *take action.* In his case, having scant evidence that the Germans were working on an atomic bomb and not convinced that the bomb's use would be necessary to end the war in Japan, Einstein wrote an urgent letter to President Roosevelt in the spring of 1945 expressing his concerns. There is no evidence that Roosevelt read it before he died in April. After bombs were dropped on Japan (during the first months of the Truman administration), Einstein was devastated and lamented how much weight his 1939 letter was said to have carried in launching the whole enterprise. When questioned about his role by *Newsweek* in 1947, he said, "Had I known that the Germans would not succeed in producing an atomic bomb...I never would have lifted one finger" (as quoted in Isaacson, 2007, p. 485). Like his professional colleagues who founded the Federation of Atomic Scientists and the *Bulletin of Atomic Scientists*, he devoted much of the rest of his life to advocating publicly for a world government and universal disarmament.

Conclusion: Where the buck stops

In all of the cases described in this chapter, creative individuals and their contributions encounter unanticipated developments; they are faced with the question of whether to act, and, if so, how strongly and in which ways. The association between the individuals and these unanticipated developments may look considerably different across cases: in some the association is actively sought and in others it comes as a surprise; in some the association bestows considerable benefits on individuals while in others it exacts considerable costs. In all of the cases, though, the response most properly resides with the creative individual themselves. In responding to a powerful and influential threat, similarly powerful and influential forces are needed; these can only come from the individual responsible for the idea or work under threat. Creative individuals are not responsible solely *for* ideas; as we have argued, they are also responsible *to* ideas, and this responsibility extends well beyond the moment at which they put their ideas into the world. Each of the artists and scientists we have considered in this chapter recognized this principle to varying degrees and relied on different resources when responding.

Those individuals who worked in an established profession, like physicist Eugene Rabinowitch or political scientist Robert Putnam, had more firmly established ethical guidelines on which to draw. Those individuals working in emerging professions or creative non-professions, like Gretchen Peters or our lightly fictionalized Picasso, needed to draw more on informal social networks and on their individual powers of discernment. In all cases, though, the "buck" stopped with the creative individual. Returning to Csikszentmihalyi's nodes, we offer this observation: although individuals may draw on the traditions of their domain and the wisdom of their peers in the field for guidance, the ultimate responsibility for creative contributions when they are threatened rests firmly and solely with the individual. Once ideas are in the world, individuals whose livelihood and reputations are often bound up in their ideas must exercise a restrained but firm grasp on how they are used. This is no easy task—more of an art than a science, one might say—but it is the ethical responsibility that accompanies creativity and innovation in all domains. In that sense, the incredible joy that accompanies a fundamental discovery or an original creation may ultimately be balanced by a fearsome responsibility.

Note

1. And there are many examples: during the 2008 presidential campaign alone, Republican candidates received cease-and-desist orders, open letters, or public chastising from John Mellencamp, Jackson Browne, Heart, Boston, Orleans, and the Foo Fighters. See http://www.rollingstone.com/music/news/stop-using-my-song-republicans-a-guide-to-disgruntled-rockers-20081010.

References

Atomic Scientists of Chicago. (1945). *Bulletin of the Atomic Scientists of Chicago, 1*(1), 1.

Bach, S. (2008). *Leni: The Life and Work of Leni Riefenstahl.* New York: Vintage Books.

Cotter, H. (2012). Alighiero Boetti Retrospective at Museum of Modern Art. *The New York Times*, June 28. http://www.nytimes.com/2012/06/29/arts/design/alighiero-boetti-retrospective-at-museum-of-modern-art.html, accessed November 6, 2013.

Csikszentmihalyi, M. (1996). *Creativity: Flow and the Psychology of Discovery and Invention.* New York: HarperCollins.

Dannen, G. (1998). Atomic Bomb: Decision—The Franck Report, June 11, 1945. http://www.dannen.com/decision/franck.html, accessed September 14, 2012.

Edgers, G. (2009). Shepard the giant. *Boston Globe*, January 25. http://www.boston.com/ae/theater_arts/articles/2009/01/25/shepard_the_giant/?page=full, accessed November 6, 2013.

Eksteins, M. (1990). *Rites of Spring: The Great War and the Birth of the Modern Age.* New York: Anchor Books.

Fairey, S. (2012). The importance of fair use and artistic freedom. *Huffington Post*, September 7. http://www.huffingtonpost.com/shepard-fairey/statement-about-my-trial_b_1864090.html, accessed November 6, 2013.

Feldman, D. H., Csikszentmihalyi, M., & Gardner, H. (1994). *Changing the World: A Framework for the Study of Creativity.* Westport, CT: Praeger.

Gardner, H. (1983). *Frames of Mind: The Theory of Multiple Intelligences.* New York: Basic Books.

Gardner, H. (1995). Reflections on multiple intelligences: Myths and messages. *Phi Delta Kappan, 77*(3), 200–03, 206–09.

Gardner, H. (2011a). *Creating Minds*, 2nd edn. New York: Basic Books.

Gardner, H. (2011b). *Truth, Beauty, and Goodness Reframed: Educating for the Virtues in the Twenty-First Century.* New York: Basic Books.

Gardner, H., Csikszentmihalyi, M., & Damon, W. (2001). *Good Work: When Excellence and Ethics Meet.* New York: Basic Books.

Isaacson, W. (2007). *Einstein: His Life and Universe.* New York: Simon and Schuster.

Jonas, M. (2007). The downside of diversity. *Boston Globe*, August 5. http://www.boston.com/news/globe/ideas/articles/2007/08/05/the_downside_of_diversity/?page=full, accessed November 6, 2013.

Kennedy, R. (2012). Shepard Fairey is fined and sentenced to probation in "Hope" poster case. *ArtsBeat*, September 7. http://artsbeat.blogs.nytimes.com/2012/09/07/shephard-fairey-is-fined-and-sentenced-to-probation-in-hope-poster-case/, accessed November 6, 2013.

Lloyd, J. (2006). Study paints bleak picture of ethnic diversity. *Financial Times*, October 8. http://www.ft.com/intl/cms/s/0/c4ac4a74-570f-11db-9110-0000779e2340.html#axzz258ahNNWw, accessed November 6, 2013.

Mcdonald, S. (2008). Yosi Sergant and the art of change: The publicist behind Shepard Fairey's Obama hope posters. *LA Weekly*, September 10. http://www.laweekly.com/2008-09-11/columns/yosi-sergant-and-the-art-of-change-the-publicist-behind-shepard-fairey-39-s-obama-hope-posters/, accessed November 6, 2013.

Price, M. (1995). Roots of dissent: The Chicago Met Lab and the origins of the Franck Report. *Isis, 86*(2), 222–244.

Putnam, R. D. (2007). E pluribus unum: Diversity and community in the twenty-first century. *Scandinavian Political Studies, 30*(2), 137–174. doi:10.1111/j.1467-9477.2007.00176.x.

Rabinowitch, E. (1952). Ten years after. *Bulletin of the Atomic Scientists, 8*(9), 294–295, 314.

Shelburne, C. (2008). "Independence Day" writer donates to Planned Parenthood in Sarah Palin's name. *CMT Blog*, October 6. http://blog.cmt.com/2008-10-06/independence-day-writer-donates-to-planned-parenthood-in-sarah-palins-name/, accessed November 6, 2013.

Shorrock, R. (2009). Spreading the hope: Street artist Shepard Fairey. *Fresh Air*, January 20. National Public Radio. http://www.npr.org/templates/story/story.php?storyId=99466584, accessed November 6, 2013.

Slaney, P. D. (2012). Eugene Rabinowitch, the Bulletin of the Atomic Scientists, and the nature of scientific internationalism in the early Cold War. *Historical Studies in the Natural Sciences, 42*(2), 114–142.

Trimborn, J. (2007). *Leni Riefenstahl: A Life*. New York: Faber and Faber.

Weber, M. (1958). Religious rejections of the world and their directions. In H. H. Gerth & C. W. Mills (eds.), *From Max Weber: Essays in Sociology* (pp. 323–359). New York: Oxford University Press.

Part II

When, How, and Why Does Creativity Lead to Positive or Negative Ethical Impacts—or Both?

The following five chapters examine how aspects of creativity—such as personality, identity, freedom and constraints of the problem, problem construction, and technology—can affect the ethics of the work process or outcome.

6
A Creative Alchemy

Ruth Richards
Saybrook University, McLean Hospital, and Harvard Medical School, USA

Can creativity change us for the better? Let us hope so, for then our job is not simply to keep some unprincipled creators in rein (these malevolent creators do indeed exist; Cropley et al., 2010), while convincing others to put more effort toward a greater good. The task instead may be to help us all unfold our deepest and most positive potential.

Some—including psychotherapists, expressive artists, spiritual teachers, and writers of diaries, day pages, and blogs—believe that the creative process is capable of opening us to a greater authenticity, presence, self-knowledge, personal integration, and access to our unconscious realms. We become more consciously aware and mindful, and less driven by habit, past experience, fears, expectations, or covert social forces. We also may become more attuned to others, more ethically driven, and better able to bypass our own ego limitations and tune into more universal themes (Dudek, 1993; Richards, 2007a, b, c, 2010; Rogers, 2011; Runco & Richards, 1998).

Humanistic psychologist Abraham Maslow (1968, 1971) foreshadowed positive psychology by studying high-functioning, self-actualizing people opening to their fullest human potential. He saw ethics as central. He even saw these individuals as representing a cutting edge of humanity, "the growing tip"; indeed, one that will "manage to flourish in a hundred or two hundred years, if we manage to endure" (Maslow, 1982, p. 168). Hence, if we humans take our next steps in hand, rather than leaving them to genetics (or blind epigenetics, or power politics, or random chance), we might help fashion, as humans, our own *conscious evolution*—our cultural evolution, both individually and together (Combs & Krippner, 1999; Maslow, 1982; Ornstein, 1991; Richards, 1998). As we will see later, Charles Darwin would have agreed.

Example: Smile Cards

You pull up at a bridge tollbooth. The tolltaker gives you a Smile Card and waves you on. What is this? The car before you has actually paid *your* toll! You look at the colorful business-sized card, with a big blue SMILE, a yellow smiley face, and a sentence in black: "You've just been tagged!" You received an anonymous act of kindness. Now it's your turn to "pay it forward." Any act will do. Will you smile?

Not just someone's wild creative idea, Smile Cards have made it around the world, an example of the gift economy, based on shifts toward contribution, trust, community, and abundance (versus concerns with grasping, distrust, isolation, and scarcity, so prevalent today). Young interns have tracked where some of the cards have gone. The sponsor, www.ServiceSpace.org, has been responsible for half a million people taking part in one act of giving or another. This includes a restaurant called Karma Kitchen, where every check reads zero, and customers are invited to "pay forward" for the person dining after them. A variant on this model is Aravind Eye Hospitals in India, where one paid client allows two poor patients to be treated for free (Mehta & Shenoy, 2011). This has become the largest vision-restoration project in the world, and it is a required case study for Harvard Business School. Is this human nature or simply one way in which people act?

Two Perspectives: Moran's two roles for creativity in a society

The current premise: Creativity, all else being equal, can make us better, more moral people. This is a shining alchemy. By *creativity*, we refer both to certain products (from work, leisure, or everyday life) showing originality and meaningfulness (Barron, 1969; Richards, 2007a) and to the creative process involved in their genesis. We touch on all four "Ps" of creativity as we proceed further: product, process, person, and press of the environment (Richards, 2007a). Everyday creativity can apply to almost anything. Plus, it is not only what we do, but how we do it. Participants in assessment studies I did with Dr. Dennis Kinney and others at Harvard Medical School included a talented auto mechanic who designed his own tools (Runco & Richards, 1998). The term *moral*—although one can debate the particulars—pertains to qualities of character and behavior and to standards for right or wrong, for virtue and just actions (American Heritage Dictionary, 2000). A greater good for a group is often involved. The term *ethics* often pertains to standards for a person or profession.

Moran identified two roles for creativity, relating to relationships that obtain "between an activity (creativity) and its environment writ large": improvement and expression (2010, p. 75).

Improvement

We focus most on the second role, expression (which we will also subdivide), but look first at improvement, where the focus is on creative product, viewed from outside, involving an end product with utility to some group. Here is our eminent or socially recognized (more than everyday) creator. From a systems perspective, we have an entire dance of people, each with a vested interest—not just the creator, but the benefactors, regulators, and finally the consumers—who snatch up the newest toothpaste to iPhone to Kindle to Netflix. There is also a vested interest in the past: this creativity is meeting some past need or goal.

Furthermore, from a systems perspective: (1) we are all open systems in an evolving whole (Moran, 2010; Richards, 2000–2001), constantly changing, with no parts fully separable; (2) our creativity seems key in human evolution (for example, Arons, 2007; Eisler, 2007; Loye, 2007; Gabora & Kaufman, 2010), becoming even more so perhaps in the future (Abraham, 2007); (3) both everyday and eminent creativity must be involved for biological and cultural evolution (also suggesting that creativity is prosocial, not harmful); and (4) consumers of creativity also will be more readily engaged or "early adopters" if they manifest more everyday creativity (Pritzker, 2007; Richards, 1998, 2010; Zausner, 2007).

Expression

Our concern is mainly with process, experience, personal change, and implications for the larger social context. The improvement role is more external, social, and objective. By contrast, our expressions are more internal, personal, and subjective (Moran, 2010). We consider not just product, but how the creative person operates, learns, and grows. This is also aligned to an as yet indefinable future where creative process can open us to new expressions, discoveries, personal growth, and sometimes group emergence. It may even change the environment. Here is the real adventure of creativity: risking the status quo to court the unknown. Might doing this be beneficial? Might it have moral implications? Those are good questions. Below, we look separately at personal and interpersonal creativity, considering examples and the issues they raise. Could creativity actually help us become better people?

Expression—Type I: Personal creativity

Four examples follow that give us a flavor for how creative product, process, and personal development may manifest. Also keep in mind the environmental press: such expressions will be rarer in some parts of the world than others. The first example may resonate for some people, whether telling a friend or therapist a secret, writing a blog, or doing expressive arts that plumb our experience. We are taking a risk.

Pennebaker's expressive writing research

In the best-known example from the Pennebaker studies (Lepore & Smyth, 2002; Pennebaker, Kiecolt-Glaser, & Glaser, 1988), college students wrote expressively over several sessions about a traumatic experience they had never shared with anyone. This involved both the expression of difficult emotions and some thoughtful processing. Meanwhile, a control group wrote about something neutral. Some participants may never have acknowledged this traumatic material to themselves, never mind shared it. One remarked: "Before...I'd lie to myself...Now...I got it off my chest" (Pennebaker, Kiecolt-Glaser, & Glaser, 1988, p. 300).

Although afterward, the expressive group members were acutely more upset, they later made fewer visits to the student health center and showed greater psychological wellbeing than the controls. Remarkably, six weeks later, they were also higher on two blood measures of immune function. Even their T-cells knew the difference! Evolution itself was perhaps voting for expressive disclosure. Expressive writing also frees working memory. Do participants become more free, integrated, whole, in touch with events that were unconscious? Are they more mindful, less apt to act out? More honest with self and others, even when difficult? More compassionate to others with similar trials? These are good questions.

Artists who open things up for us

Some artists, whether eminent or everyday creators, will not wait for us to ask them. They share new truths, including ones we do not want to hear. An artist friend speaks of "canaries in a coal mine." For some, this is a highly moral calling.

Stefanie Dudek spoke of Transgressive Art and the moral urgency of expression, of shaking up the status quo, of finding and sharing a higher truth. She noted how Henry Miller said, "Art consists in going the full length so that by the emotional release those who are dead may be restored to life" (Dudek, 1993, p. 149). Or how composer John Cage

called on "morality if you wish...to intensify, alter perceptual awareness and, hence, consciousness." Or Davey, on art: "It is entirely itself.
It is not deferring to external value systems. It carries a primary level
meaning. Meanwhile... [witness the integrity of the artist]... it subverts
cliché, exposes subtle distortions" (Dudek, 1993, p. 147).
No outside moral enforcer needed; values are intrinsic to this work.
Expression is not just for the artist; it is for the culture. Think of Picasso's
painting *Guernica*, a stark black-and-white newsreel portrayal of the horrors of war with, center front and poignant, one terrified horse. The
truth leads. Is this so for all of art? Some of it? For science? Are the
fidelity and bravery of the artist intrinsically moral? Again, these are
good questions.

Self-actualizing people

Consider the impacts over time of, in effect, an everyday creative
lifestyle. Maslow (1968, 1971; also see Richards, 2007a, b) came to see
little difference, in his exceptional participants marked by a high level
of psychological development, when comparing self-actualizing qualities generally and self-actualizing creativity. These were also really good
people, contributing creatively in the world.

Maslow found special qualities revealed in the ordinary doings of
life, such as openness, humor, and approaching almost anything creatively. His creators lived more "in the real world than in the verbalized
world of concepts and abstractions. There was less blocking, less self-
criticism, more spontaneity." As with "the creativeness of all secure and
happy children," it was "spontaneous, effortless, innocent, easy, a kind
of freedom from stereotypes and clichés" (Maslow, 1968, p. 88). Yet, it
was not so simple; these individuals were also more whole, integrated,
non-defensive, not self-conscious, committed to serious work, yet able
to have fun. Maslow noted: "The civil war within the average person
between the forces of the inner depths and the forces of defense and
control seem to have been resolved in my subjects and they are less
split... more of themselves is available for use, for enjoyment, and for
creative purposes" (1968, p. 90).

These creators had very strong egos (in the "adaptation to reality sense": Barron, 1969), yet were also "very easily egoless, and
self-transcending" (Maslow, 1968, p. 89). Often, work and play were
related, with work centering on something of benefit and not so much
on themselves alone. What once might have been "deficiency creativity" (to defend, or to solve one's problems) had moved toward
"being creativity," a higher expression of one's humanness (Rhodes,
1998).

Part of this was commitment to *being values*, for example, truth, justice, beauty, and—of real interest—aliveness (Maslow, 1971). Here is a bow toward conscious awareness and living more actively in the present moment. Maslow (1970) later wished he had paid more attention to creative people in groups. We will get there soon.

Compensatory advantage

This type of expressive creativity may seem contradictory since its origin is in major psychiatric disorders. But—surprise!—it, too, is about health. The concept of "compensatory advantage" is often misunderstood. It is fine to be a little off center, to "stand apart" (Richards & Kinney, 1989), when one creatively seeks a new view and throws a quick curve to the status quo.

Everyday creativity is linked to milder psychopathology, if any. It favors better-functioning individuals in the bipolar or schizophrenia spectrum, or more seriously ill people during their better-functioning states. Thus, both trait and state phenomena are relevant, as well as both personal and family psychiatric history. "Normal" relatives also may show a creative edge (Kinney et al., 2000–2001; Richards et al., 1988; see also Silvia & Kaufman, 2010). Significantly, the bipolar and schizophrenia spectra disorders are not rare: by one estimate, over 5 percent of the general population may have bipolar spectrum disorder (Runco & Richards, 1998).

Yet, if creativity, for some people, is "one of their best medicines," how tragic when some romanticize illness and resist highly helpful treatments. A full one out of five (or 20 percent) of people with bipolar I disorder who do *not* get treatment actually take their own lives (Jamison, 1993). It is bad enough when health or creativity is lost; here, lives are lost.

Some individuals with personal or family bipolar risk find benefits from which we all can learn, involving richer associations, emotions, confidence, and extra energy. Somewhat different compensatory advantages may exist for schizophrenia (Schuldberg & Sass, 2000–2001). Either way, a balance of inspiration and control is needed, as in Kris's (1950; in Richards, 1981) well-known "regression in the service of the ego." One can view this through many lenses, including the difference between schizophrenic patients with little manifest creativity who have low latent inhibition (LI), versus high-achiever, creative types who show low LI along with psychological controls. Looser gating may allow good ideas, but requires adaptive oversight. Low LI has been linked to sound health and qualities including high creativity, openness to experience (Carson, Peterson, & Higgins, 2003), and "faith in intuition" (Kaufman & Singer, 2012). This is a complex balance.

Does a creative "compensatory advantage" help maintain personal or familial genetic diatheses for major psychiatric illnesses strongly in the population? We postulate a morally positive direction to evolution. Some have further posited more altruistic motives among individuals in the bipolar spectrum (Goodwin & Jamison, 2000; Runco & Richards, 1998).

Three issues: Personal creative expression

Our "creative deep"

The unconscious mind is key, whether in arts, science, or self-actualizing creative living (Hassin, Uleman, & Bargh, 2005; Jung, 1983; Richards, 1981, 2007a, b; Russ & Fiorelli, 2010; Zausner, 2007). Can we reach and manage it? Witness comedian Robin Williams on free association from the television series *Mork and Mindy*: "You are in control but you're not—the characters are coming through you" (Russ, 1993, p. 21). Here is flexible control plus authenticity in an emergent performance. Kaufman and Singer's (2012) reframing of Daniel Kahneman's fast and slow types of thinking resonates here. Russ's (1993) work suggests—for cognition and affect—that use of primitive material in childhood play, in conjunction with a parallel, more reality-oriented processing, can predict both adult creativity and resilience. The current proposition: As we become more open, aware, spontaneous, non-defensive, and integrated, our creativity and better human nature may both come forward, as with Maslow's (1968, 1971) self-actualizers.

Creative deep is a descriptive name for this rich repository of hidden experience and memory, myths, magic, images, symbols, metaphors, "reflectaphors," worlds of fantasy. It also contains fear, including closeted stores of painful or "shadow" material we cannot face, humming beneath the surface, dynamic and alive (Briggs, 1987; Jung, 1983; Richards, 2007a, 2010; Singer & Sincoff, 1990). Witness inventor Nikola Tesla setting a machine running in his mind and testing it later for wear (Tart, 1975). Nor do phenomena concern only the personal unconscious, but also the collective unconscious (Jung, 1983).

So huge is this hidden realm of mind, we can almost say we don't know ourselves. Similar to the metaphorical iceberg, where 10 percent sticks out of the water, we look in the mirror and think we see ourselves, but it is only the 10 percent, with a whole 90 percent unseen. Sometimes this material, still masked, may emerge in painful pathology. Hence, for many psychotherapies, a key goal is making the unconscious conscious (Richards, 1981).

Our neurobiology and altered states of consciousness (ASC)

It is curious how many people see creativity as just another aspect of ordinary waking mind. But if we look at Wallas's (1926) stages of the creative process—preparation, incubation, intimation, illumination, verification—the middle three, at minimum, are far from ordinary mind. (*Intimation*, the one often omitted, is the sense that "something is coming.") Notably, Martindale (1999) discovered low cortical activation, dominant right-hemisphere activity, and low frontal activity including slow theta waves during receptive creating, or incubation. The concurrent subjective experience? Defocused attention, associative thought, and multiple representations, all occurring simultaneously. We find this slow theta wave activity elsewhere, including in meditation (Austin, 1998; Richards, 2007b). We know meditation can improve us in varied ways, opening a path to peace, wellbeing, and health writ large, for mind, body, and spirit (Richards, 2007a, b).

Neurophenomenological studies using EEG and/or fMRI support the relevance of cognitive disinhibition to creativity (Kaufman et al., 2010) while confirming the presence of slow waves and right-sided activity during receptive insight-generating states. How interesting that "insight" strategies for solving certain creative word problems shows this pattern, but "analytical" strategies do not. Some people can even "turn on" this mind state ahead of time (Subramanian et al. and Kounios et al., cited in Richards, 2010).

Remarkable! Creativity can alter our consciousness. With practice, might we even do so at will? *Altered states of consciousness* (ASC) are defined as "stable patterns of physiological, cognitive, and experiential events different from those of the ordinary waking state" (Baruss, 2003, p. 8). Perhaps forms of writer's block come from our inner critic kicking us out of a magical ASC and back into our hyper-self-critical conceptual beta-waves (Richards, 2007a, b). Might more spacious ASCs help us see more deeply to compassion and a greater good? Many spiritual leaders would agree (Baruss, 2003; Dalai Lama, 1999; Nhat Hanh, 1975; Richards, 2007a).

Creative normalcy

Refreshing? A *creative* norm? Not the same as our social norm, which is sometimes conformist and mindless. For creativity, let us also honor not only the mean, but the standard deviation. Might some people fear a norm so open, authentic, real, present, and dangerous—acknowledging our depths, honoring the irrational? Possibly. Would it be easier to project it onto some marginal group? Surely, such groups are available with the link to psychopathology (Richards, 1996, 1998, 2010). If such "projective identification" exists, it is to everyone's discredit. We have proposed "broadening the acceptable limits of normality" for creativity (Runco & Richards, 1998) to honor our intrapersonal depth and interpersonal variation, our sparkling and healthy human diversity so vital in this shrinking world of globalization. Our norms should be culturally competent, aware, dynamically evolving, and surely not pathologizing those who reject a mindless, lock-step sameness. Will a moral high ground lie with some of our open, authentic, and tolerant creators? It is a good question.

Unfortunately, what if a "crazy" stereotype sometimes carries over to us, the creators, and we are viewed as a little absent-minded, oddly dressed, hair uncombed, socially inept, odd of speech, bumping into walls? Eysenck (1993) found "psychoticism" to be linked with creativity, but "a little" psychoticism is only colorful, and not at all psychotic; see Schuldberg's "normal" college students (Schuldberg 1998). Try doing a Google Images search for "mad creator." Is that you? It is doubtful.

Beyond psychiatric issues, recall Barron's (1969) studies of eminent writers who scored high on the Minnesota Multiphasic Personality Inventory (MMPI) scales. These writers were exceptionally high on ego strength. Perhaps they had access to a more colorful "creative palette" and knew how to use it. Drawing from chaos theory, Krippner, Richards, and Abraham (2011) suggested that new and creative "strange attractors" of mind are more likely with an optimal balance of divergent and convergent thinking. Again, this evokes Kris's "regression in the service of the ego." Recall Barron's famous quote: "The creative person is both more primitive and more cultivated, more destructive and more constructive, occasionally crazier and yet adamantly saner, than the average person" (1963, p. 234).

We should highlight the "adamantly saner" above. If creativity links with illness, why do we find its healthy benefits everywhere, yielding stress resistance, resilience, empowerment, and other benefits for physical and psychological wellbeing (Richards, 2007a, 2010; Rogers, 2011; Runco & Richards, 1998)? Why do we have healing arts, expressive arts therapies, and more, for depression, HIV, cancer, bereavement, or trauma (Richards, 2010; Rogers, 2011)? Why do we find creativity linked, by humanistic psychologists, with the highest health and levels of moral development? Then there is humility: a moral plus. Even without an Eastern view of creativity (Sundararajan & Averill, 2007), we who create can become, consciously at times, supplicants: we risk the unknown, give up control. We do not pridefully say this is "my unique idea"—the Western stereotype—or at least not at the time. We plumb our hidden psychological depths. This not only reveals new creative possibilities but can help change us, the supplicants. We are patient, receptive, open. Our brains are on slow waves. We are even open to contradictions with our cherished views of self. We value something higher. Humbled and in awe, we are open to vast realms that we, by definition, know nothing about. We risk the new. Where might wisdom (for example, Moran, 2010) intertwine with creativity? Perhaps here. If more people accept such norms, will we also be more open to others? Not necessarily. Some do not prioritize this. Yet, when it happens, it can be extraordinary.

Expression—Type II: Interpersonal creativity

Interpersonal or relational creativity—here is where product can meet process in a dance of relationship. This is discussed further elsewhere (Richards, 2007b, c; Sawyer, 2010; and see Eisler, 2007; Goerner, 2007; Loye, 2007). We note briefly here two cases involving dyads and groups. If we define interactional creativity using originality and meaningfulness, we may quickly differentiate between conversations A and B. In A, someone is recounting his or her achievements; another is not listening ("uh huh, uh huh"). No points here for creativity (or for human connection). In conversation B, there is rapt attention, listening, spontaneous responding, and interchange, even risking of self in a willingness to hear and learn and change. As per Buber's (1970) *I and Thou*, there may be the emergence of something more, deep and powerful, a "between space." It comes alive. Creativity and values are intertwined. Two examples follow.

Dyads

In dissertations at Saybrook University on creative relationships, in psychotherapy and creativity, participants emerge present, empathetic, and flexible. Relationships are intimate and infused with commitment to honesty, trust, collaboration, and moving past blocks to a higher truth. Consider empathy from Relational Cultural Theory (Jordan et al., 1991) at Wellesley College and McLean Hospital, Harvard (where I worked clinically on an experimental inpatient unit using this theory): client and therapist are direct, present, aware, non-hierarchical, and open to change in response to the other. Criteria of originality and meaningfulness characterize the product, which is also process (Richards, 2007c). One knows what can happen without such trust, directness, and creative commitment.

Groups and organizational change

One may take interpersonal creativity to a group level, involving stages of consciousness and emergent phenomena, which may even pull us "forward into the future." Otto Scharmer (2009) at MIT developed these prospects in *Theory U*. To simplify from a complex business model, we take an example of a physician dialoguing with a terminally ill patient and his or her family. The commitment to engage is a moral commitment to dialogue creatively, hear the other, and honor the process. Four stages may occur, moving successively into greater openness, authenticity, risking, and group awareness, toward emergence and new vision. Scharmer sees the fourth stage, "generative systems," as "the central phenomenon of our time" (2009, p. 361). These stages are:

- *Autistic:* no real listening, doctor and family expound, nothing new gets in.
- *Adaptive*: some new material registers; here is the beginning of communication with the doctor, including fears of death.
- *Self-reflective:* self is part of larger picture, helping the system see itself; it dawns on people that "we are the system" transforming into emergent plans.
- *Generative:* "connecting with the deepest presence and future possibility"; from stilted discussion to honest appraisal of fears, caring, life expectancy, with co-evolution into what people can realistically do for a beloved parent.

How does the doctor (as the group leader) manage this? The leader, far from "being driven by past patterns and exterior forces," is able to "shift the inner place from which we operate" (Scharmer, 2009, p. 373), while engaging others in its definition. In chaos theory terms, this resembles a shift, where a new group *attractor* becomes defined and dominant (Briggs & Peat, 1989; Richards, 2000–2001). It is like leaving the orbit of Earth and approaching Venus close enough to fall within its gravitational field, and the group co-creates Venus.

Scharmer's organizational process includes meditation, particularly when courting levels of openness to solutions before group emergence: "Retreat and reflect, allow the inner knowing to emerge" (2009, p. 378). Powerful practices foster ASCs and conditions for humility, listening, presence, engagement, and intuitive resonance to solutions toward a greater good—a high moral and creative activity.

Two issues: Interpersonal creative expression

The lost Darwin

Yes, humans do cooperate. How wrong some had it about Darwin. Only the strong survive? Nature is "red in tooth and claw"? Not always for humans. That was Herbert Spencer and Social Darwinism. Darwin, who, incidentally, once trained for the ministry, had a kinder view. In his later writings, he spoke to human collaboration, morality, and sympathy (similar to empathy). In fact, in *Descent of Man* he used "survival of the fittest" only twice, once to apologize for the term. By contrast, he used the word "love" 95 times (Loye, 2007). We humans are a species that can and should work together. Darwin saw this as part of our collective future and higher calling, a natural and moral evolution for human beings specifically. Our empathy in pairs and our larger emergent patterns in groups may herald higher levels of both creativity and moral development (see Eisler, 2007; Goerner, 2007; Richards, 2007b).

Chaos theory and emergence

Chaos theory and non-linear dynamical systems theory, with their breathtaking fractal forms of nature ("fingerprints" of "strange attractors" that resonate throughout living/non-living realms, marking growth and change in our world), are altering our understanding. From a linear cause-and-effect model of separate independent entities, we move to a more holistic, complex, deeply interconnected, and at times unpredictable systems model. It is non-linear because, unlike Math 101, twice the push does not always mean twice the result. At times, nothing happens. At other times, when an organism/system is "sensitive to initial conditions" and far from an energetic equilibrium—like a creative mind poised at the top of a mental cliff awaiting the key datum—a small push can create an avalanche to a new attractor (example of the Butterfly Effect). A thunderstorm is a classic example and, metaphorically at least, so is a creative insight (Briggs & Peat, 1989; Richards, 2000–2001).

The term *chaos* refers not to anarchy, but to deep underlying order. At times, we find, remarkably, that order simply emerges (the whole is greater than the sum of its parts). From the first molecules of life to hierarchical systems and natural ecologies, potentially including human ones, many systems seem to emerge. They self-organize for enhanced adaptability, creativity, and complexity. Is that not amazing?

Consider the proposition that states of consciousness, including ASCs, are related to discrete attractors showing self-organization (Combs & Krippner, 1999, 2007). Through a sudden shift or "bifurcation" from one attractor to another, minds can shift near instantaneously to new holistic states: for example, from waking to sleep. There is a sudden onset and a totally different end state. Even more intriguing is when diverse instances fit within a few deeper underlying patterns. Consider the fact that there are 250,000 types of plants, yet only three forms of distribution of leaves around stems, with the spiral accounting for 80 percent (Combs & Krippner, 1999). Are there distinct emergent stages in how we know ourselves and our world?

Combs and Krippner draw on Gebser's historical "structures of consciousness" (stable structures underlying our worldviews) said to reflect "a progression of ways humans come to understand reality" (2007, p. 136) from early humanity to the present. These structures include archaic, magical, mythic, and mental. They organize our states of consciousness. Some believe the integral structure is now becoming more prominent, with a more fluid, multimodal, multiperspectival, and creative capacity.

Might stage 4 of Scharmer's (2009) *Theory U* herald a new emergence for an evolving human future? A state, certainly. What about a structure or worldview? A step toward more evolved humanity? Speculative, yes, but Scharmer's (2009) group process holds something new, requiring attunement and advanced moral development. Individual uniqueness remains valued, yet within an open, creative group honoring an emergent future. Perhaps you have witnessed such remarkable shifts in groups. They can bring chills. Magic moments also occur between couples. Darwin would have approved.

Again, Smile Cards

How do these cards, or the Gift Economy, connect? Actually rather well, from personal consciousness-raising to global involvement, from creative product to process to even a changing person, to hope of changing our cultural values, the press of the environment. Smile Cards may not be for everyone, but one likely can empathize with the greater vision and imagine a different subculture emerging. We may even seek our own Creativity of Giving. How hopeful when, like Aravind, it becomes an international movement.

Is something here contagious? Do we feel better, more worthwhile, more compassionate, part of a larger system—one that cares about us, too, and lacks the cruelty of the rest of the world? It sounds great. But is a Gift Economy realistic? Buddha, Lao Tzu, Jesus, and Mohammed thought so. Yet, only time will tell.

Conclusion

Is being creative and living creatively—all else being equal—not only intrinsically healthy but potentially transformative? Can this take us to a more open, authentic, aware, non-defensive, attuned, connected, caring, and morally elevated place, where we can live better, individually and together? There is initial support here, but we still need more research.

The three systems perspectives on creative roles above involved (a) improvements (with an external focus on eminent creators, creative product, and societal criteria), plus, the main focus here, two process-oriented perspectives on (b) individual expressions (with an internal focus on everyday creativity and processes catalyzing personal change) and (c) interpersonal expressions (with an internal focus on process as product through emergent group phenomena). Five related issues with moral implications were considered: our creative deep, altered states of consciousness, creative "normalcy," the Lost Darwin (surprisingly prosocial), and chaos theory and emergence. Creativity, so viewed, could bring health and humility that could deepen our awareness, responsiveness to others, and ability to co-create—and to relate—in troubled times. It could even advance our conscious evolution (Ornstein, 1991).

Yet, what makes us believe that we, as mere humans, are good enough to carry this off? Are we angels or are we demons? Joseph Campbell addressed this question extensively through mythic views of human potential, including twelfth- and thirteenth-century Arthurian Romance. Here, one sees the hope of the human "who in his native virtue is competent both to experience and to render blessedness, even in the mixed field of this our life on earth" (1964, p. 507). Campbell frames the best of human possibility as an ongoing quest across world wisdom traditions, including Christian, Celtic, Islamic, Byzantine, Buddhist, and many others: "Take the mystery of the Grail: For what reason, pray, should a Christian knight ride forth...when at hand, in every chapel...[is] the sacrament...The answer, obviously, is that the Grail Quest was an individual adventure in experience" (1964, p. 508). Campbell shows that the Grail Hero is not a great person, but the "forthright, simple, uncorrupted, noble" offspring of nature, "not an angel or a saint" (1964, p. 508), but a questing individual guided by courage, compassion, loyalty, and steadfastness. Here, then, we also may find the ultimate alchemy, an homage to the "open heart," the lesson of the eternal human awakened, "and the icons of our common heritage: Microcosm and macrocosm are in essence one in God" (1964, pp. 516–517).

We are neither angels nor saints. But can we find in our human creativity our own alchemy and a path to our higher possibilities? That is the hope, and the reason for this chapter.

References

Abraham, F. D. (2007). Cyborgs, cyberspace, cybersexuality: The evolution of everyday creativity. In R. Richards (ed.), *Everyday Creativity and New Views of Human Nature* (pp. 241–259). Washington, DC: American Psychological Association.

American Heritage Dictionary, 4th edn. (2000). Boston, MA: Houghton Mifflin.

Arons, M. (2007). Standing up for humanity: Upright body, creative instability, and spiritual balance. In R. Richards (ed.), *Everyday Creativity and New Views of Human Nature* (pp. 175–193). Washington, DC: American Psychological Association.

Austin, J. H. (1998). *Zen and the Brain*. Cambridge, MA: MIT Press.

Barron, F. (1963). *Creativity and Psychological Health*. Princeton, NJ: Van Nostrand.

Barron, F. (1969). *Creative Person and Creative Process*. New York: Holt, Rinehart & Winston.

Baruss, I. (2003). *Alterations of Consciousness*. Washington, DC: American Psychological Association.

Briggs, J. (1987). Reflectaphors: The (implicate) universe as a work of art. In B. Hiley & F. D. Peat (eds.), *Quantum Implications* (pp. 414–435). London: Routledge.

Briggs, J., & Peat, F. D. (1989). *Turbulent Mirror*. New York: Harper & Row.

Buber, M. (1970). *I and Thou*. New York: Touchstone/Simon and Schuster.

Campbell, J. (1964). *The Masks of God*. New York: Viking Compass.

Carson, S., Peterson, J. B., & Higgins, D. M. (2003). Decreased latent inhibition is associated with increased creative achievement in high functioning individuals. *Journal of Personality and Social Psychology, 85*(3), 499–506.

Combs, A. L., & Krippner, S. (1999). Spiritual growth and the evolution of consciousness. *International Journal of Transpersonal Studies, 18*(1), 11–21.

Combs, A. L. & Krippner, S. (2007). Structures of consciousness and creativity. In R. Richards (ed.) *Everyday Creativity and New Views of Human Nature* (pp. 131–149). Washington, DC: American Psychological Association.

Cropley, D., Cropley, A., Kaufman, J. C., & Runco, M. A. (2010). *The Dark Side of Creativity*. New York: Cambridge University Press.

134   *The Ethics of Creativity*

H. H. Dalai Lama (1999). *Ethics for the New Millennium*. New York: Riverhead Books.

Dudek, S. Z. (1993). The morality of 20th-century transgressive art. *Creativity Research Journal, 6*(1–2), 145–152.

Eisler, R. (2007). Our great creative challenge: Rethinking human nature and recreating society. In R. Richards (ed.), *Everyday Creativity and New Views of Human Nature* (pp. 261–285). Washington, DC: American Psychological Association.

Eysenck, H. (1993). Creativity and personality: Suggestions for a theory. *Psychological Inquiry, 4*, 147–178.

Gabora, L., & Kaufman, S. B. (2010). Evolutionary approaches to creativity. In J. Kaufman & R. Sternberg (eds.), *Cambridge Handbook of Creativity* (pp. 279–300). New York: Cambridge University Press.

Goerner, S. (2007). A "knowledge ecology" view of creativity: How integral science recasts collective creativity as a basis of large-scale learning. In R. Richards (ed.), *Everyday Creativity and New Views of Human Nature* (pp. 221–239). Washington, DC: American Psychological Association.

Goodwin, F. K., & Jamison, K. R. (2000). *Manic-Depressive Illness*. New York: Oxford University Press.

Hassin, R. R., Uleman, J. S., & Bargh, J. A. (2005). *The New Unconscious*. New York: Oxford University Press.

Jamison, K. R. (1993). *Touched with Fire*. New York: Free Press.

Jordan, J., Kaplan, A. G., Miller, J. B., Stiver, I. P., & Surrey, J. L. (1991). *Women's Growth in Connection*. New York: Guilford Press.

Jung, C. G. (1983). *The Essential Jung* (A. Storr, ed.). Princeton, NJ: Princeton University Press.

Kaufman, A. B., Kornilov, S., Bristol, A., Tan, M., & Grigorenko, E. (2010). The neurobiological foundation of creative cognition. In J. Kaufman & R. Sternberg (eds.), *Cambridge Handbook of Creativity* (pp. 216–232). New York: Cambridge University Press.

Kaufman, S. B., & Singer, J. L. (2012). The creativity of dual process "system 1" thinking. http://blogs.scientificamerican.com/guest-blog/2012/01/17, accessed February 10, 2013.

Kinney, D. K., Richards, R., Lowing, P., LeBlanc, D., Zimbalist, M., & Harian, P. (2000–2001). Creativity in offspring of schizophrenics and controls. *Creativity Research Journal, 11*, 17–25.

Krippner, S., Richards, R., & Abraham, F. D. (2011). Creativity and chaos in waking and dreaming states. *NeuroQuantology, 10*(2), pp. 164–176.

Lepore, S. J., & Smyth, J. M. (2002). *The Writing Cure: How Expressive Writing Promotes Health and Emotional Well-Being*. Washington, DC: American Psychological Association.

Loye, D. (2007). Telling the new story: Darwin, evolution, and creativity versus conformity in science. In R. Richards (ed.), *Everyday Creativity and New Views of Human Nature* (pp. 153–174). Washington, DC: American Psychological Association.

Martindale, C. (1999). Biological bases of creativity. In R. Sternberg (ed.), *Handbook of Creativity* (pp. 137–152). New York: Cambridge University Press.

Maslow, A. (1968). Creativity in self-actualizing people. In A. Rothenberg & C. A. Hausman (eds.), *The Creativity Question* (pp. 86–92). Durham, NC: Duke University Press.

Maslow, A. (1970). *Religions, Values, and Peak-Experiences*. New York: Viking Compass.

Maslow, A. (1971). *The Farther Reaches of Human Nature*. New York: Penguin Books.

Maslow, A. (1982). *The Journals of Abraham Maslow* (R. Lowry, ed.). Brattleboro, VT: Lewis Publishing Company.

Mehta, P. K., & Shenoy, S. (2011). *Infinite Vision: How Aravind Became the World's Greatest Business Case for Compassion*. San Francisco, CA: Berrett-Koehler.

Moran, S. (2010). The roles of creativity in society. In J. Kaufman & R. Sternberg (eds.), *Cambridge Handbook of Creativity* (pp. 74–90). New York: Cambridge University Press.

Nhat Hanh, T. (1975). *Miracle of Mindfulness*. Berkeley, CA: Parallax Press.

Ornstein, R. (1991). *The Evolution of Consciousness*. New York: Prentice Hall.

Pennebaker, J. W., Kiecolt-Glaser, J. K., & Glaser, R. (1989). Disclosure of traumas and immune function. In M. Runco & R. Richards (eds.), *Eminent Creativity, Everyday Creativity, and Health* (pp. 287–302). Greenwich, CT: Ablex.

Pritzker, S. (2007). Audience flow. In. R. Richards (ed.), *Everyday Creativity and New Views of Human Nature* (pp. 109–130). Washington, DC: American Psychological Association.

Rhodes, C. (1998). Growth from deficiency creativity to being creativity. In M. Runco & R. Richards (eds.), *Eminent Creativity, Everyday Creativity, and Health*. Greenwich, CT: Ablex.

Richards, R. (1981). Relationships between creativity and psychopathology: An evaluation and interpretation of the evidence. *Genetic Psychology Monographs, 103*, 261–324.

Richards, R. (1996). Does the lone genius ride again? Creativity, chaos, and community. *Journal of Humanistic Psychology, 36*(2), 44–60.

Richards, R. (1998). When illness yields creativity. In M. Runco & R. Richards (eds.), *Eminent Creativity, Everyday Creativity, and Health* (pp. 485–540). Greenwich, CT: Ablex.

Richards, R. (2000–2001). Millennium as opportunity: Chaos, creativity, and J. P. Guilford's Structure-of-Intellect Model. *Creativity Research Journal, 13*(3–4), 249–265.

Richards, R. (ed.). (2007a). *Everyday Creativity and New Views of Human Nature*. Washington, DC: American Psychological Association.

Richards, R. (2007b). Twelve potential benefits of living more creatively. In R. Richards (ed.), *Everyday Creativity and New Views of Human Nature* (pp. 261–286). Washington, DC: American Psychological Association.

Richards, R. (2007c). Relational creativity and healing potential: Power of Eastern thought in Western clinical settings. In J. Pappas, W. Smythe, & A. Baydala (eds.), *Cultural Healing and Belief Systems* (pp. 286–308). Calgary: Detselig.

Richards, R. (2010). Everyday creativity: Process and way of life—Four key issues. In J. Kaufman & R. Sternberg (eds.), *Cambridge Handbook of Creativity* (pp. 189–215). New York: Cambridge University Press.

Richards, R., & Kinney, D. K. (1989). Creativity and manic-depressive illness (letter). *Comprehensive Psychiatry, 30*, 272–273.

Richards, R., Kinney, D. K., Lunde, I., Benet, M., & Merzel, A. (1988). Creativity in manic-depressives, cyclothymes, their normal relatives, and control subjects. *Journal of Abnormal Psychology, 97*, 281–288.

Rogers, N. (2011). *The Creative Connection® for Groups: Person-Centered Expressive Arts for Healing and Social Change*. Palo Alto, CA: Science & Behavior Books.

Runco, M. & Richards, R. (eds.) (1998). *Eminent Creativity, Everyday Creativity, and Health*. Greenwich, CT: Ablex.

Russ, S. W. (1993). *Affect and Creativity: The Role of Affect and Play in the Creative Process*. Hillsdale, NJ: Lawrence Erlbaum Associates.

Russ, S., & Fiorelli, J. A. (2010). Developmental approaches to creativity. In J. Kaufman & R. Sternberg (eds.), *Cambridge Handbook of Creativity* (pp. 233–249). New York: Cambridge University Press.

Sawyer, R. K. (2010). Individual and group creativity. In J. Kaufman & R. Sternberg (eds.), *Cambridge Handbook of Creativity* (pp. 366–380). New York: Cambridge University Press.

Scharmer, C. O. (2009). *Theory U: Leading from the Future as It Emerges*. San Francisco, CA: Berrett-Koehler.

Schuldberg, D. (1998). Schizotypal and hypomanic traits, creativity, and psychological health. In M. Runco & R. Richards, R. (eds.), *Eminent Creativity, Everyday Creativity, and Health* (pp. 157–172). Greenwich, CT: Ablex.

Schuldberg, D., & Sass, L. (eds.) (2000–2001) Special issue: Creativity and the schizophrenia spectrum, *Creativity Research Journal*, 13(1).

Silvia, P., & Kaufman, J. C. (2010). Creativity and mental illness. In J. Kaufman & R. Sternberg (eds.), *Cambridge Handbook of Creativity* (pp. 381–394). New York: Cambridge University Press.

Singer, J., & Sincoff, J. B. (1990). Summary: Beyond repression and the defenses. *Repression and Dissociation* (pp. 471–496). Chicago, IL: University of Chicago Press.

Sundararajan, L., & Averill, J. (2007). Creativity in the everyday: Culture, self, and emotions. In R. Richards (ed.), *Everyday Creativity and New Views of Human Nature* (pp. 195–219). Washington, DC: American Psychological Association.

Tart, C. T. (1975). *States of Consciousness*. New York: Dutton & Co.

Wallas, G. (1926). *The Art of Thought*. New York: Harcourt Brace.

Zausner, T. (2007). Artist and audience: Everyday creativity and visual art. In R. Richards (ed.), *Everyday Creativity and New Views of Human Nature* (pp. 75–90). Washington, DC: American Psychological Association.

7
License to Steal: How the Creative Identity Entitles Dishonesty

Lynne C. Vincent
Vanderbilt University, USA

and

Jack A. Goncalo
Cornell University, USA

Organizations operate in an increasingly uncertain and changing world. Competition on the domestic and international fronts is intense, and organizations must create new products, strategies, services, and methods for maneuvering in the changing environment. As a result, organizations are recognizing the value of employees' creativity as a way to innovate and maintain a competitive advantage (Thompson, 2003). Researchers and organizations are now beginning to explore how having a creative identity can increase creativity (Farmer, Tierney, & Kung-McIntyre, 2003; Jaussi, Randel, & Dionne, 2007). Given the power of identities for shaping performance outcomes (Ashforth & Mael, 1989; Beyer & Hannah, 2002; Wrzesniewski, Dutton, & Debebe, 2003), it is not surprising that recent research has begun to explore how the creative identity can also motivate creative behavior.

The creative identity (Farmer, Tierney, & Kung-McIntyre, 2003; Jaussi, Randel, & Dionne, 2007) is the overall importance a person places on being creative as part of his or her self-definition (Jaussi, Randel, & Dionne, 2007). Consistent with other important social or personal identities (Aquino & Reed, 2002; Cheryan & Bodenhausen, 2000), the creative identity can contribute to a self-fulfilling prophecy whereby individuals who see themselves as creative are motivated to engage in behaviors that support that positive self-concept as a creative individual (Petkus, 1996). Interestingly, these behaviors may include objectively creative behaviors such as generating new ideas (Jaussi, Randel, & Dionne, 2007), but they may also include behavior that the individual

137

may associate with creativity but that may not always facilitate creative performance (Petkus, 1996; Wells, 1978). It is the latter possibility that is the focus of this chapter.

Here, we delve into the dark side of the creative identity. Although there is emerging evidence that a creative identity can motivate creative achievement, we argue that identifying oneself as creative might also cause individuals to develop an exaggerated sense of entitlement stemming from that fact that creativity is typically viewed as a rare and valuable attribute (Campbell et al., 2004). The consequences of entitlement may be wide-ranging and potentially damaging. We focus specifically on dishonesty and argue that a strong sense of entitlement will lead to dishonest behavior, because individuals feel that they justly deserve more than others and are willing to engage in unethical behaviors to gain those rewards. Given the increasing importance of creativity within organizations and the costs of dishonesty for those organizations, understanding why individuals with creative identities are prone to dishonesty could be crucial for organizations that desire creative performance. By understanding the psychological consequences of a creative identity, both positive and negative, managers may be able to strengthen employees' creative identities without incurring the costs.

The creativity identity: Antecedents and consequences

Existing research has viewed creativity primarily as a cognitive ability or a personality trait (see Hennessy & Amabile, 2010 for a review). Creativity as a self-concept is a newer construct that has emerged only recently. As such, the creative identity holds a great deal of promise but is not yet well understood.

In general, a salient identity provides individuals with a frame of reference that allows them to evaluate how their actions fit with social expectations (Wells, 1978). They are motivated to align their behavior with role expectations as a way of protecting a positive self-view and projecting that view to others. In theory, the creative identity should operate through a similar process. Individuals may derive considerable positive meaning and value from a salient self-view as a creative person (Lemons, 2010; Petkus, 1996). Therefore, individuals with creative identities will engage in behaviors that they view as creative (Petkus, 1996) in order to maintain and protect this highly positive self-view (Fisher, 1997; Petkus, 1996). For these individuals, creativity represents a central part of "who they are" (Farmer, Tierney, & Kung-McIntyre, 2003).

Like other identities, the creative identity may be more salient in some situations than in others (Farmer, Tierney, & Kung-McIntyre, 2003). Creativity-relevant norms (Ford, 1996), supervisors' expectations (Scott & Bruce, 1994), and co-worker support and interaction (Amabile et al., 1996; Zhou & George, 2001) can cause an individual's creative identity to move to the fore. An identity provides a set of internalized expectations for behavior. As normative expectations of important "social others" are a major source of an individual's self-concept through reflexivity, or seeing oneself through such expectations, others' expectations can cause an identity to develop (for example, Callero, Howard, & Piliavin, 1987) and cause individuals to engage in behaviors that support that identity (Ford, 1996). For example, in a study of Taiwanese employees, Farmer and colleagues (2003) demonstrated that when employees perceived that co-workers expected them to be creative, their role identities as creative employees were stronger. And, similar to the way other valued role identities are maintained, individuals may engage in behaviors that they perceive as creative in order to preserve and affirm a positive self-image as a creative individual (Farmer, Tierney, & Kung-McIntyre, 2003; Jaussi, Randel, & Dionne, 2007; Lemons, 2010; Petkus, 1996).

What is particularly intriguing, however, is the way in which a creative identity differs from creative ability. That is, merely having a creative identity does not necessarily guarantee that an individual is objectively more creative than others (Lemons, 2010). An individual can have a creative identity without having pronounced creative ability. Rather, the creative identity merely provides the motivation to engage in behaviors that are viewed as creative (Petkus, 1996). As such, individuals with creative identities are more likely to engage in creative behaviors at work (Jaussi, Randel, & Dionne, 2007). Jaussi and colleagues argue that the creative identity motivates individuals to engage consistently in creative behaviors because creativity is fundamental to their self-definition. Those individuals want to be creative and will engage in behaviors to support that positive self-image. While individuals with creative identities tend to have some creative ability (Lemons, 2010) and believe that they can be effectively creative (Jaussi, Randel, & Dionne, 2007), possessing a creative identity helps to explain why individuals seek out opportunities to be creative beyond creative ability and creative self-efficacy (Jaussi, Randel, & Dionne, 2007).

There is a great deal of room for error, however, if the creative identity includes behaviors that are stereotypically associated with creative individuals but do not necessarily contribute to creative performance.

Indeed, there is the rather perverse possibility that individuals may attempt to support a creative identity by enacting behaviors that may in fact impede the creative process or even have unintended negative consequences for other performance outcomes. We consider this intriguing possibility in the next section.

The creative identity and psychological entitlement

Building and strengthening a creative personal identity form a promising avenue through which organizations can stimulate creative performance (Farmer, Tierney, & Kung-MacIntyre, 2003). The logic behind this assertion is that because the creative identity is valuable, people will be motivated to maintain this self-image by actively seeking out opportunities to demonstrate creative achievement (Jaussi, Randel, & Dionne, 2007; Petkus, 1996). It may be premature to adopt this optimistic view, however, before we have a comprehensive understanding of the repertoire of behaviors that people may associate with the creative identity. The characteristics associated with creative people range from social desirable to social negative and may include the tendency to be self-interested, independent, non-conforming, and unconventional (Feist, 1998; Gough, 1979).

Creative people are expected to be confident risk-takers who defy the status quo to move the group in a new direction. Thus, they may be viewed as exceptional and given wide latitude to deviate from expectations. It is telling that a recent national survey of CEOs conducted in the United States showed that creativity is the most desired trait in an effective leader (Mueller, Goncalo, & Kamdar, 2011). In many organizations, creative ideas are viewed as an important source of competitive advantage; in society more generally, creativity is heralded as an engine of progress. Creative individuals are rewarded with resources, status, and recognition while their less creative counterparts work in their shadows (Audia & Goncalo, 2007).

Individuals who are seen as creative can receive more freedom because of this reputation (Baucus et al., 2008). For example, technological innovator Steve Jobs had a habit of parking his Mercedes in handicap parking spots and driving it without a license plate (Isaacson, 2011). His biographer Isaacson noted, "I think he felt the normal rules just shouldn't apply to him" (Messick, 2011). By virtue of their creativity, creative individuals are allowed to transgress society's norms and, at times, they are lauded for doing so. This leniency is commonly seen in organizations that value creativity. Baucus et al. (2008) reviewed the four most common prescriptions for how organizations promote creativity, which include: (1) breaking the rules or avoiding standard approaches to problems; (2) challenging authority and avoiding traditions; (3) welcoming conflict; and (4) taking risks. Such leniency may inadvertently signal that creative people are unique and deserve special treatment. The extreme value placed on creative achievement may shape the way people experience and enact the creative identity. Viewing oneself as a creative individual may, in addition to fueling a sense of uniqueness, also generate a heightened sense of entitlement based on the belief that, as a creative individual, one is contributing something of great value and importance (Goncalo, Flynn, & Kim, 2010).

The value that organizations place on creativity may moderate the extent to which the creative identity leads to entitlement. An individual's personal identity is a multifaceted combination of self-concepts (McCall & Simmons, 1966; Monin & Jordan, 2009), and individuals' behaviors will change as different facets of the self-concept become salient (Aquino & Reed, 2002; Hart, Atkins, & Ford, 1998). As identity is malleable, certain situations will cause different facets of an individual's identity to become salient (Farmer, Tierney, & Kung-McIntyre, 2003). For instance, when organizations encourage creativity in their employees by creating a norm to be creative or supervisors positively recognize an employee for their creativity, the creative identity may become salient. When individuals perceive that an identity is valued, they are more likely to engage in behaviors associated with that identity (McCall & Simmons, 1966). Therefore, the relationship between the creative identity and entitlement may be affected by the value that the organization or the field places on creativity (Csikszentmihalyi, 1996). For instance, in companies such as innovation firms or Hollywood studios that value creativity, the relationship between the creative identity and entitlement may be increased. However, in organizations that do not value creativity explicitly, that same relationship may be decreased.

How the creative identity entitles dishonesty

If individuals with creative identities feel entitled, this belief could affect their behaviors. Specifically, this belief could cause them to engage in dishonesty. Entitled individuals "prefer being treated as special or unique in social settings" (Snow, Kern, & Curlette, 2001, p. 104) and expect outcomes and events to "go their way" (2001, p. 106). Psychological entitlement can increase selfishness and reduce helping behaviors (Zitek et al., 2010). For instance, individuals who scored higher on the Psychological Entitlement Scale took more candy from a bowl that was to be shared with sick children in hospitals, said they deserved higher salaries than other workers, acted more selfishly in a common dilemma game, and treated their romantic partners in a more selfish manner (Campbell et al., 2004). Entitled individuals often want to receive more output for the same level of input as others, and they will seek out or create situations where they are receiving more than average (Huseman, Hatfield, & Miles, 1987).

A creative identity may make people feel entitled to more than they truly deserve. Here, we take this logic one step further to suggest that a salient creative identity may not only make people feel entitled to more, but they may also feel entitled to steal in order to tip the scale in their favor. Because entitled individuals feel as if they justly deserve more than others, a sense of entitlement allows individuals to rationalize or reframe dishonest behaviors (Mazar & Ariely, 2006) through moral disengagement (Bandura, 1999). For instance, an entitled individual may view theft as merely claiming what they justly deserve. Because the creative identity causes individuals to feel special and entitled, they feel that they deserve more. This exaggerated feeling of entitlement provides them with an excuse to steal with impunity.

It is important to note how our argument relating the creative identity to dishonesty differs from related work on creative thinking. For example, Gino and Ariely (2012) found that creative thinkers are cognitively flexible and are better able to justify or rationalize their dishonest behavior. Similarly, research has also shown that positive affect can promote dishonesty, by making people more cognitively flexible and thus more adept at justifying and rationalizing their behavior (Vincent, Emich, & Goncalo, 2013). In sum, existing research has focused on creative ability to argue that cognitive flexibility may lead to moral flexibility. Thus, existing research linking creativity to dishonesty implicates the unintended negative consequences of creative thought. In contrast, we argue that the creative identity (as opposed to creative personality or creative thinking) can also facilitate dishonesty, not because the creative identity confers the ability to think more flexibly, but because it entitles people to take more than they deserve.

Reducing dishonesty: The perils of self-awareness

Dishonesty such as lying, cheating, or stealing is a common and costly issue for organizations: 95 percent of all companies have reported experiencing difficulties regarding dishonest behaviors within their organization (Henle, Giacalone, & Jurkiewicz, 2005). These difficulties are not the results of the misdeeds of a few employees, the proverbial "bad apples." Rather, the majority of employees tend to be dishonest to some extent. Up to 75 percent of employees have engaged in some form of theft, fraud, embezzlement, vandalism, sabotage, or unexcused absenteeism (Harper, 1990; US Mutual Association, 1998). More than half of these employees will engage in theft repeatedly (US Mutual Association, 1998). The small misdeeds of many employees lead to staggering costs for organizations: a typical organization loses approximately 5 percent of its revenue to fraud annually, resulting in a global loss of 2.9 trillion dollars (Association of Certified Fraud Examiners, 2010).

Given the significant costs of dishonesty, researchers have investigated methods of reducing dishonesty in organizations. One of the most consistent and robust findings in this area demonstrates that increasing self-awareness increases honesty by enhancing conformity to salient rules and norms, which generally include norms of morality (Carver, 2003). If discrepancies between oneself and the salient internal standard are detected, the self-aware person will either alter their behavior to conform to the salient internal standards of behavior or alter the components of the standard (for example, Carver, 1974, 2003; Scheier, Fenigstein, & Buss, 1975).

By increasing one's self-awareness, dishonesty is often reduced (Carver, 2003). For instance, children trick-or-treating on Halloween were more likely to follow the rules and less likely to steal (that is, take only one piece of candy) when the standard of taking one piece of candy was made salient and when the children were individualized by being asked their names and addresses or when a mirror was placed directly behind the candy bowl (Beaman et al., 1979). Similarly, college students were less likely to cheat on a task when they performed that task in front of a mirror (Diener & Wallbom, 1976). By increasing self-awareness, individuals focused their attention on themselves and the salient standard, which, in this case, was to be honest (for example, by taking one piece of candy or by not cheating on a task). Retail stores employ a similar logic when using surveillance cameras to decrease rates of theft. The presence of the surveillance cameras reminds individuals of the norm to be honest and also of the consequences of being caught performing the dishonest acts. On being reminded, most individuals will conform to the standard.

Yet, for individuals with creative identities, increasing self-awareness may act in a counterintuitive way. As the foregoing review would suggest, for individuals with a creative identity, feeling entitled, unique, and special may be the more likely self-view to be enhanced when they are made to be self-aware. In other words, for individuals with a creative identity, self-awareness could activate the very feelings of entitlement that we theorize will lead to dishonest behavior. Self-awareness affects narcissists in a similar manner by enhancing a salient standard. Similar to entitled individuals, narcissists tend to have an inflated sense of self-importance and to self-enhance their own performance (Westen, 1990). When performing tasks in front of a mirror to increase self-awareness, narcissists' tendency to self-enhance grew, their self-confidence increased, and they overestimated their contribution to a group project (Robins & John, 1997). However, the opposite effect occurred for non-narcissists, who reduced their tendency to self-enhance and reported lower self-confidence (Robins & John, 1997). As self-awareness increases narcissists' tendency to self-enhance, self-awareness could enhance a sense of entitlement for individuals with a creative identity. Thus, the creative identity provides a behavioral standard that may cause individuals to feel even more entitled, which may cause them to engage in dishonesty. Therefore, contrary to previous findings regarding self-awareness, heightened self-awareness for individuals with creative identities may increase rather than decrease the tendency for those individuals to engage in dishonest behavior by increasing the sense of entitlement.

The creative identity and other counterproductive behaviors

If individuals with a salient creative identity believe that they deserve special treatment, what happens when they do not receive the special attention they feel they deserve? This is likely to happen given that the creative identity may motivate people to attempt creative work although they are not guaranteed success in creative endeavors. Indeed, many organizations will claim to want creative ideas, but creative ideas may be dismissed in favor of more practical solutions (Mueller, Melwani, & Goncalo, 2012; Staw, 1995). Creative ideas are novel by definition and may initially generate conflict and controversy, particularly if they threaten to upend the status quo (Nemeth & Staw, 1989). Moreover, while many organizations claim to want creative leadership, there is evidence that people who express creative ideas are actually perceived to have less leadership potential compared to employees who express more practical ideas (Mueller, Goncalo, & Kamdar, 2011). People who attempt to be creative may expect to be lauded for their efforts, but most will endure rejection (Goncalo, Vincent, & Audia, 2010). It is not too surprising that the creative process can be a source of frustration for most people; however, for an entitled individual, the reaction could be significantly more negative, aggressive, and severe. Research suggests that entitled narcissists are easily offended (McCullough et al., 2003; Witte, Callahan, & Perez-Lopez, 2002) and are more likely to respond negatively to a message from a competitor (Exline et al., 2004). Moreover, entitled individuals are less likely to forgive but are more likely to insist on repayment for an offense (Bishop & Lane, 2002; Exline et al., 2004). Furthermore, feeling victimized can trigger a sense of entitlement (Zitek et al., 2010), so feeling victimized or offended may cause the entitled individual's negative reaction to the perceived offense or slight to be exacerbated.

Beyond being less likely to forgive, entitled individuals are also more likely to be aggressive after being criticized. Campbell and colleagues (2004) found that entitled individuals who had been criticized behaved aggressively against their criticizer. After receiving a negative evaluation on a short essay that they had written, entitled individuals chose a longer and more severe punishment for their criticizer. Campbell et al. (2004) argue that criticism contradicts entitled individuals' feelings of deservingness. When this self-concept as special and unique is threatened, the entitled individual can become more aggressive. Moreover, as entitlement causes a focus on the self and a neglect of the concerns of others, entitled individuals perceive aggression toward the criticizer as a reasonable response (Campbell et al., 2004).

In sum, when the creative identity is salient, individuals may more easily be offended and react more negatively to criticisms regarding their work. In particular, if their identity as creative is criticized, the entitled individuals may react negatively and potentially negatively against the criticizer.

We began with the optimistic view that the creative identity could motivate creative achievement. However, increased performance on a creative task may not translate to other kinds of tasks. Indeed, the creative identity may actually diminish performance on tasks that do not demand creative solutions. Time is a valuable resource for every person. However, as the entitled individual believes that they are more valuable than other people and should receive special treatment, they might think that their time is more valuable than other people's time. While everyone must complete some dull or tedious tasks at work, entitled individuals may perceive these tasks as a waste of their valuable time (O'Brien, Anastasio, & Bushman, 2011). Moreover, during dull tasks, entitled individuals believe that more time has passed (O'Brien,

Anastasio, & Bushman, 2011). In other words, time drags when entitled individuals are not having fun.

As a result, entitled individuals may become bored more easily, dissatisfied with their job, or even resentful of having to complete these tasks. They may react to this frustration in several ways. First, they may be insulted that they were even asked to complete the task. They then may become aggressive toward the individual or individuals they associate with these tasks. They may engage in dishonest behavior as a way to get revenge against the organization that is causing them to suffer. Second, they may not complete the tasks at all and choose to spend their time in other ways. While they may be productive in those chosen tasks, not completing the dull tasks could cause difficulties for the organization or other employees. Third, the individual may become so dissatisfied with the situation that they remove themselves from it, causing the organization to incur the costs of replacing the employee.

Individuals with creative identities may be particularly affected by dull tasks or those that do not promote their creative identity. As the creative identity may represent a core component of their self-concept, those individuals may be particularly sensitive to tasks that do not require or allow them to be creative. Moreover, they may feel that these tasks are a waste of their time. As such, individuals with creative identities may require more sensation or variation in tasks in order to remain satisfied with a job and focused on a task.

Interventions that may reduce dishonesty

If the creative identity does promote a sense of entitlement that, in turn, causes dishonest behaviors, selfish behaviors, reduced helping behaviors, and even aggression, the value of promoting a creative identity becomes questionable. By promoting a creative identity, organizations may incur the unintended cost of promoting these negative interpersonal behaviors. While the benefits of creativity are significant, the costs of the negative interpersonal behaviors cannot be ignored. The costs of dishonesty alone are overwhelming. Moreover, reducing these negative behaviors could be difficult as increasing self-awareness, which is a typical method of increasing honest behaviors, could backfire by increasing the salience of the feeling of entitlement.

Given this, how can organizations encourage employees to have a creative identity without causing them to feel more entitled? Rather than increase self-awareness, organizations should intervene by directly addressing the aspect of the creative identity that causes dishonest behavior: entitlement. Organizations could address the issue of entitlement in multiple ways. First, they could attempt to lower employees' sense of entitlement. This could be done by asking employees to consider their connections to others and why they do not necessarily deserve more than others. However, this request would be need to be repeated in order to be effective, which may make it difficult and inefficient for organizations to implement.

Second, organizations could also try to create a group creative identity rather than encourage individual employees' creative identities. Entitlement is a very self-focused trait. Entitled individuals are more concerned for their own outcomes than the outcomes of others (Huseman, Hatfield, & Miles, 1987). However, by focusing on a collective identity as creative, the self-focus may be reduced and shift toward a focus on the collective. As a result, the sense of entitlement may be reduced while maintaining the creative identity. In this way, organizations may be able to receive the benefits of the creative identity without the costs of increased dishonest behaviors. To develop an organizational creative identity, organizations could encourage creativity by explicitly mentioning it in formal and informal communications or mission statements (Lee et al., 2004), using creative role models (Shalley & Perry-Smith, 2001; Zhou, 2003), requiring employees to be creative as a formal job requirement (Unsworth, Wall, & Carter, 2005), or by providing employees with creativity goals (Shalley, 1991, 1995).

Conclusion

In this chapter, we have argued that the creative identity has an unexplored dark side. On one hand, the creative identity promotes creative behaviors, which could help organizations generate new products and ideas. However, this creative identity also triggers a sense of entitlement, which, in turn, causes the individual to engage in dishonest behaviors. This relationship should not be ignored, as organizations are increasingly encouraging their employees to be creative. The connection between creativity and dishonesty extends beyond the creative personality and the creative ability to be morally flexible. This research demonstrates that the mere self-perception as creative is sufficient to motivate dishonesty. While creative ability allows individuals to justify their dishonest deeds, the relationship between creative identity and dishonesty does not appear to function through moral flexibility. Rather, this chapter argues that the creative identity includes a sense of entitlement and that people who perceive themselves as creative are more likely to engage in dishonest behaviors. The creative identity is a relatively new and underdeveloped construct. By understanding this construct, organizations may be able to achieve the notable and significant benefits of creativity while reducing the unintended but potentially significant costs of employee dishonesty.

References

Amabile, T. M., Conti, R., Coon, H., Lazenby, J., & Herron, M. (1996). Assessing the work environment for creativity. *Academy of Management Journal, 39*, 1154–1184.

Aquino, K., & Reed, A. I. I. (2002). The self-importance of moral identity. *Journal of Personality and Social Psychology, 83*, 1423–1440.

Ashforth, B. E., & Mael, F. (1989). Social identity theory and the organization. *Academy of Management Review, 14*, 20–39.

Association of Certified Fraud Examiners. (2010). *2010 ACFE Report to the Nation on Occupational Fraud & Abuse.* http://www.acfe.com/uploadedFiles/ACFE_Website/Content/documents/rttn-2010.pdf, accessed April 2, 2012.

Audia, P. G., & Goncalo, J. A. (2007). Success and creativity over time: A study of inventors in the hard-disk drive industry. *Management Science, 53*, 1–15.

Bandura, A. (1999). Moral disengagement in the perpetration of inhumanities. *Personality and Social Psychology Review, 3*, 193–209.

Baucus, M. S., Norton, W. I., Jr., Baucus, D. A., & Human, S. E. (2008). Fostering creativity and innovation without encouraging unethical behavior. *Journal of Business Ethics, 81*, 97–115.

Beaman, A. L., Klentz, B., Diener, E., & Svanum, S. (1979). Self-awareness and transgression in children: Two field studies. *Journal of Personality and Social Psychology, 37*(10), 1835–1846.

Beyer, J. M., & Hannah, D. R. (2002). Building on the past: Enacting established personal identities in a new work setting. *Organization Science, 13*, 636–652.

Bishop, J., & Lane, R. C. (2002). The dynamics and dangers of entitlement. *Psychoanalytic Psychology, 19*, 739–758.

Callero, P. L., Howard, J. A., & Piliavin, J. A. (1987). Helping behavior as role behavior: Disclosing social structure and history in the analysis of prosocial action. *Social Psychology Quarterly, 50*, 247–256.

Campbell, W. K., Bonacci, A. M., Shelton, J., Exline, J. J., & Bushman, B. J. (2004). Psychological entitlement: Interpersonal consequences and validation of a self-report measure. *Journal of Personality Assessment, 83*, 29–45.

Carver, C. S. (1974). Facilitation of physical aggression through objective self-awareness. *Journal of Experimental Social Psychology, 10*, 365–370.

Carver, C. S. (2003). Self-awareness. In M. R. Leary & J. P. Tangney (eds.), *Handbook of Self and Identity* (pp. 179–196). New York: Guilford.

Cheryan, S., & Bodenhausen, G. V. (2000). When positive stereotypes threaten intellectual performance: The psychological hazards of model minority status. *Psychological Science, 11*, 399–402.

Csikszentmihalyi, M. (1996). *Creativity: Flow and the Psychology of Discovery and Invention.* New York: HarperCollins.

Diener, E., & Wallbom, M. (1976). Effects of self-awareness on anti-normative behavior. *Journal of Research in Personality, 10,* 107–111.

Exline, J. J., Baumeister, R. F., Bushman, B. J., Campbell, W. K., & Finkel, E. J. (2004). Too proud to let go: Narcissistic entitlement as a barrier to forgiveness. *Journal of Personality and Social Psychology, 87,* 894–912.

Farmer, S., Tierney, P., & Kung-McIntyre, K. (2003). Employee creativity in Taiwan: An application of role identity theory. *Academy of Management Journal, 46,* 618–630.

Feist, G. J. (1998). A meta-analysis of personality in scientific and artistic creativity. *Personality and Social Psychology Review, 2*(4), 290–309.

Fisher, T. (1997). The designer's self-identity: Myths of creativity and the management of teams. *Creativity and Innovation Management, 6*(1), 10–18.

Ford, C. (1996). A theory of individual creative action in multiple social domains. *Academy of Management Review, 21,* 1112–1142.

Gino, F., & Ariely, D. (2012). The dark side of creativity: Original thinkers can be more dishonest. *Journal of Personality and Social Psychology, 102*(3), 445–459.

Goncalo, J. A., Flynn, F. J., & Kim, S. H. (2010). Are two narcissists better than one? The link between narcissism, perceived creativity, and creative performance. *Personality and Social Psychology Bulletin, 36,* 1484–1495.

Goncalo, J. A., Vincent, L. C., & Audia, P. G. (2010). Early creativity as a constraint on future achievement. In D. Cropley, J. Kaufman, A. Cropley, & M. Runco (eds.), *The Dark Side of Creativity* (pp. 114–133). New York: Cambridge University Press.

Gough, H. G. (1979). A creative personality scale for the adjective check list. *Journal of Personality and Social Psychology, 37,* 1398–1405.

Harper, D. (1990). Spotlight abuse. Save profits. *Industrial Distribution, 79,* 47–51.

Hart, D., Atkins, R., & Ford, D. (1998). Urban America as a context for the development of moral identity in adolescence. *Journal of Social Issues, 54,* 513–530.

Henle, C. A., Giacalone, R. A., & Jurkiewicz, C. L. (2005). The role of ethical ideology in workplace deviance. *Journal of Business Ethics, 56*(3), 219–230.

Hennessey, B. A., & Amabile, T. M. (2010). Creativity. *Annual Review of Psychology, 61,* 569–598.

Huseman, R. C., Hatfield, J. D., & Miles, E.W. (1987). A new perspective on equity theory: The equity sensitivity construct. *Academy of Management Review, 12,* 222–234.

Isaacson, W. (2011). *Steve Jobs.* New York: Simon and Schuster.

Jaussi, K. S., Randel, A. E., & Dionne, S. D. (2007). I am, I think I can, and I do: The role of personal identity, self-efficacy, and cross-application of experiences in creativity at work. *Creativity Research Journal, 19*(2), 247–258.

Lee, F., Edmondson, A. C., Thomke, S., & Worline, M. (2004). The mixed effects of inconsistency on experimentation in organizations. *Organization Science, 15*(3), 310–326.

Lemons, G. (2010). Bar drinks, rugas, and gay pride parades: Is creative behavior a function of creative self-efficacy? *Creativity Research Journal, 22*(2), 151–161.

Mazar, N., & Ariely, D. (2006). Dishonesty in everyday life and its policy implications. *Journal of Public Policy & Marketing, 25,* 1–21.

McCall, G. J., & Simmons, J. L. (1966). *Identities and interactions.* New York: Free Press.

McCullough, M. E., Emmons, R. A., Kilpatrick, S. D., & Mooney, C. N. (2003). Narcissists as "victims": The role of narcissism in the perception of transgressions. *Personality and Social Psychology Bulletin, 29*, 885–893.

Messick, G. (2011). *60 Minutes* (Television broadcast). October 23. New York: CBS.

Monin, B., & Jordan, A. H. (2009). Dynamic moral identity: A social psychological perspective. In D. Narvaez & D. Lapsley (eds.), *Personality, Identity, and Character: Explorations in Moral Psychology* (pp. 341–354). Cambridge: University Press.

Mueller, J. S., Goncalo, J. A., & Kamdar, D. (2011). Recognizing creative leadership: Can creative idea expression negatively relate to perceptions of leadership potential? *Journal of Experimental Social Psychology, 47*, 494–498.

Mueller, J. S., Melwani, S., & Goncalo, J. A. (2012). The bias against creativity: Why people desire but reject creative ideas. *Psychological Science, 23*(1), 13–17.

Nemeth, C. J., & Staw, B. M. (1989). The tradeoffs of social control and innovation in small groups and organizations. In L. Berkowitz (ed.), *Advances in Experimental Social Psychology* (pp. 175–210). New York: Academic Press.

O'Brien, E., Anastasio, P. A., & Bushman, B. J. (2011). Time crawls when you're not having fun: Feeling entitled makes dull tasks drag on. *Personality and Social Psychology Bulletin, 37*, 1287–1296.

Petkus, E. (1996). The creative identity. *Journal of Creative Behavior, 30*, 188–196.

Robins, R. W., & John, O. P. (1997). Effects of visual perspective and narcissism on self-perception: Is seeing believing? *Psychological Science, 8*(1), 37–42.

Scheier, M. F., Fenigstein, A., & Buss, A. H. (1975). Self awareness and physical aggression. *Journal of Experimental Social Psychology, 10*, 264–273.

Scott, S. G., & Bruce, R. A. (1994). Determinants of innovative behavior: A path model of individual innovation in the workplace. *Academy of Management Journal, 37*, 580–607.

Shalley, C. E. (1991). Effects of productivity goals, creativity goals, and personal discretion on individual creativity. *Journal of Applied Psychology, 76*, 179–185.

Shalley, C. E. (1995). Effects of coaction, expected evaluation, and goal setting on creativity and productivity. *Academy of Management Journal, 38*, 483–503.

Shalley, C. E., & Perry-Smith, J. E. (2001). Effects of social-psychological factors on creative performance: The role of informational and controlling expected evaluation and modeling experience. *Organizational Behavior and Human Decision Processes, 84*, 1–22.

Snow, J. N., Kern, R. M., & Curlette, W. L. (2001). Identifying personality traits associated with attrition in systematic training for effective parenting groups. *The Family Journal: Counseling and Therapy for Couples and Families, 9*, 102–108.

Staw, B. M. (1995). Why no one really wants creativity. In C. M. Ford & D. A. Gioia (eds.), *Creative Action in Organizations: Ivory Tower Visions and Real World Voices* (pp. 161–166). Thousand Oaks, CA: Sage.

Thompson, L. (2003). Improving the creativity of organizational work groups. *Academy of Management Executive, 17*, 96–109.

Unsworth, K. L., Wall, T. D., & Carter, A. J. (2005). Creativity requirement: A neglected construct in the field of employee creativity. *Group & Organization Management, 30*(5), 541–560.

US Mutual Association. (1998). The cost of employee dishonesty. http://www.usmutual.com/cost.html, accessed September 28, 2010.

Vincent, L. C., Emich, K. J., Goncalo, J. A. (2013). Stretching the moral gray zone: Positive affect, moral disengagement and dishonesty. *Psychological Science,* 24(4), 595–599.

Wells, L. E. (1978). Theories of deviance and the self-concept. *Social Psychology,* 41, 189–204.

Westen, D. (1990). The relations among narcissism, egocentrism, self-concept, and self-esteem: Experimental, clinical, and theoretical considerations. *Psychoanalysis and Contemporary Thought, 13,* 183–239.

Witte, T. H., Callahan, K. L., & Perez-Lopez, M. (2002). Narcissism and anger: An exploration of underlying correlates. *Psychological Reports, 90,* 871–875.

Wrzesniewski, A., Dutton, J. E., & Debebe, G. (2003). Interpersonal sensemaking and the meaning of work. *Research in Organizational Behavior, 25,* 93–135.

Zhou, J. (2003). When the presence of creative coworkers is related to creativity: Role of supervisor close monitoring, developmental feedback, and creative personality. *Journal of Applied Psychology, 88,* 413–422.

Zhou, J., & George, J. M. (2001). When job dissatisfaction leads to creativity: Encouraging the expression of voice. *Academy of Management Journal, 44,* 682–696.

Zitek, E. M., Jordan, A. H., Monin, B., & Leach, F. R. (2010). Victim entitlement to behave selfishly. *Journal of Personality and Social Psychology, 98,* 245–255.

8
Engineering, Ethics, and Creativity: N'er the Twain Shall Meet?

David H. Cropley
University of South Australia, Australia

Creativity and the role of ethics

Since Guilford (1950) first suggested divergent thinking as a component of human cognition, creativity has been associated with positive qualities, characteristics, and outcomes. On the one hand, Bruner (1962) saw creativity as a defining characteristic of human intelligence, distinguishing it from impersonal, cold, machine intelligence. A. J. Cropley drew together these positive views, which he suggested see creativity as "a principle of nature and that it is, by definition, a universal beneficial force fostering growth and rebuilding in all organic systems" (2010a, p. 2). In parallel with this view, creativity also has long been seen as good for the individual, and it is associated with many positive personal properties, such as courage, openness to experience, and flexibility, as well as offering beneficial effects for mental health (Cropley, 1990). Adding to these positive perspectives, creativity is understood to be vital culturally and organizationally "for shaping... future orientations and actualizing reforms in political, economic and cultural areas" (Oral, 2006, p. 65) and economically "as the key to meeting challenges... arising from technological advances, social change, globalization, and now the global financial crisis" (Cropley, 2010a, p. 3). In fact, whether creativity is defined and studied in relation to the person, the process, the product, or the press (environment)—namely, the 4Ps (Rhodes, 1961)—a great deal of attention is paid to the positive aspects of creativity.

More recent attention has been given to the contrasting *dark side* of creativity to shift focus to *why* most creativity is good (Cropley et al., 2010; Cropley, Kaufman, & Cropley, 2008). If malevolent creativity is "creativity that is deliberately planned to damage others" (Cropley et al., 2008, p. 106), why is a very significant proportion *not* like this? Cropley (2010b) grouped personal properties, including feelings and motivation, under the banner of *intent*, to explain why some creative endeavors are malevolent while others are benevolent. It is not difficult to see that an individual's ethics and morals play a role in setting the direction that this intent takes. Equally, the nature of the environment, or press, within which creativity takes place also influences the outcome. An organization's framework of ethics and morals— an aspect of its culture—can reinforce or diminish the individual's intent.

The purpose of this chapter is to examine the role that ethics and morals, both individual and organizational, play in shaping creativity, in particular in engineering design. While that purpose and focus are predominantly benevolent, the possibility of bad outcomes, whether accidental (negative creativity) or deliberate (malevolent creativity), should not be overlooked. Indeed, in discussing individual and organizational ethics and morals in engineering design, it is inevitable that cases such as exploding automobile fuel tanks (for example, the Ford Pinto) and collapsing hotel walkways (for example, the Kansas City Hyatt) will enter the discussion and blur the lines between benevolent creativity and malevolent creativity. What role, if any, did individual and organizational ethics and morals play in these outcomes? Are they examples of benevolent creativity gone wrong, or deliberately malevolent creativity? Regardless of how we classify them, what role did ethics and morals play in the path from the first identification of a technological problem to the implementation of an engineering solution?

Ethics, creativity, and design

The ethical framework necessary to steer creative efforts in the right direction generates a paradox that forms the key idea of this chapter. Put simply, creativity requires freedom, openness, and flexibility, whereas ethical behavior implies constraint, limitation, and control. How can the former flourish in the presence of the latter? The two appear, on the surface, to be mutually exclusive.

A simple example of the paradoxical role of ethics in creativity can be found in the case of the artist Andres Serrano's infamous "Piss Christ." If Serrano had felt bound by a code of ethics that dictated that artists should not offend the religious sensibilities of the public, then it is likely that he would not have created his notorious, and highly novel, work. Unfettered by such an ethical/moral constraint, Serrano's art was able to explore a much larger range of ideas that encompassed such novel possibilities as "immerse a religious icon in a glass of your own urine." The real or perceived absence of an ethical constraint allowed Serrano to maximize his creativity. Even if we argue that a real ethical rule was broken by the artist (as opposed to what we might think of as a softer "guideline"), we cannot deny that his willingness to overlook this constraint had only a degree of negative consequences for Serrano. His art was, in at least one case, defaced during an exhibition (Silvia & Brown, 2007) and he was accused of blasphemy (Cropley & Cropley, 2013). Even for those who were offended by it, we can argue that the negative consequences were relatively innocuous: nobody was killed or injured as a result of the production of this work of art.

Conversely, it could be argued that the artist himself has been richly rewarded for his willingness to break rules and deviate from accepted norms. Many saw it as daring, paradigm-breaking, and highly effective (Cropley & Cropley, 2013), and Serrano won a visual arts award for the work. Silvia and Brown (2007) offer some insight into why Serrano's piece might be viewed as creative. Citing Martindale and Moore (1988), they explain that people exhibit a preference for art that is prototypical for its category (let us assume, in this case, that the category is religious iconography). When a work is *not* prototypical for its category, people exhibit negative reactions. Non-prototypicality can be seen, therefore, as synonymous with non-conformity and deviance from norms, qualities that are good indicators of novelty and, thus, creativity. In simple terms, the adverse reaction of many viewers can be taken as an indicator of the creativity of Serrano's work. The point of this example is that minimizing or eliminating a constraint, or at least ignoring it, resulted in higher creativity.

By way of contrast, it is difficult to imagine the same kind of scenario playing out in a technological domain such as engineering design. Engineers generally will ignore the prevailing rules and standards of behavior only at their peril. Not only does the engineer risk personal sanction by breaking these rules, but the likelihood of negative consequences resulting from uncontrolled novelty is greater. The case of the collapsed walkways at the Kansas City Hyatt, in 1980, is a case in point. The engineer of record for the design of the hotel was charged with negligence, incompetence, and misconduct, and ultimately lost his license to practice engineering—all because of a failure to check a design change and assess its risk. While the design change that was made can be looked at as a novel solution to a particular problem, it was not effective, as demonstrated by the collapse. Cropley and Cropley (2005) have argued that it is the *combination* of novelty and effectiveness that characterizes engineering creativity.

Therefore, while Serrano's artistic creativity, built largely on a foundation of novelty, exhibited positive rewards and few negative consequences, engineering creativity, with a strong dependence on effectiveness and usually a strong coupling to human users, is much more tightly constrained. In terms familiar to researchers working with human subjects, the potential for harm resulting from engineering products is much greater than that resulting from artistic products. In engineering, this idea of potential for harm is more commonly expressed in terms of *risk*. It is hard to see the designer of a breast implant, for example, being lauded by the public as a *bold innovator* for her willingness to move outside of the ethical and safe design space and use, for example, pebbles encased in an old gym sock as an implant, simply because this is novel. Even a cursory assessment of risk would rule out this solution, no matter how novel.

In particular, in engineering, there is a fundamental tension between creativity and ethics. Any ethical framework—any set of rules governing human conduct—imposes constraints on what can and cannot be done in any given field of endeavor. Where that field of endeavor is the "production of a *perceptible product* that is both *novel and useful* as defined within a *social context*" (Plucker, Beghetto, & Dow, 2004, p. 49)—that is, creativity—then it seems self-evident that ethics will limit what can be produced. Where those products and the social context carry any potential for harm, as is usually the case in engineering, the ethical considerations drive a process of risk assessment and analysis that places severe constraints on what can and cannot be done.

It is useful, at this point in the discussion of engineering creativity, to make a distinction between what is legal and what is ethical. Cropley and Cropley (2013) cite work by Salcedo-Albarán et al. (2009) describing the differences between *statutory* rules (laws) and *customary* rules (norms). Creativity, of course, involves rule-breaking, and rules constrain creativity. However, when the rules that are broken are statutory in nature, we are more likely to perceive the product as illegal or criminal, whereas breaking customary rules is more likely to be seen as positive and creative. The design *novelty* in the case of the Kansas City Hyatt walkways was not, in itself, unethical or illegal. It was perfectly reasonable to *consider* alternative ways of supporting the walkways. The failure was in the application of a framework of ethical behavior—checking design changes, assessing their risk, certifying them as safe. Ethics does not tell us what is statutory and what is customary, but instead how we incorporate this knowledge into our behaviors and actions.

Engineering design as creativity

The paradox that has been described—creativity requires freedom, but takes place within a framework of rules that are inherently constraining—now can be examined in the specific case of engineering design. While it is true to say that engineering encompasses many varied activities, an essential core—indeed, a defining characteristic—of engineering is *design*. Dieter and Schmidt remind us that "it is true that the professional practice of engineering is largely concerned with design; it is often said that design is the essence of engineering" (2012, p. 1). Citing Blumrich (1970), they characterize the process of design as "to pull together something new or to arrange existing things in a new way to satisfy a recognized need of society" (2012, p. 1). Dieter and Schmidt (2012) describe the essence of design as *synthesis*.

Horenstein contrasted design with other essential activities in engineering by focusing on the process of solving problems:

> If only one answer to a problem exists, and finding it merely involved putting together the pieces of the puzzle, then the activity is probably *analysis*... if more than one solution exists, and if deciding upon a suitable path demands being creative, making choices, performing tests, iterating and evaluating, then the activity is most certainly *design*. Design can include analysis, but it must also involve at least one of these latter elements. (2002, p. 23)

The core of engineering practice is design, but that design activity involves two stages: a stage of creative *synthesis* followed by a stage of logical *analysis*. The first stage is synonymous with divergent thinking (Guilford, 1950), whereas the second is synonymous with convergent thinking. This may be illustrated as in Figure 8.1. Expressed in this way, we can see that engineering design is simply a form of creative problem-solving.

Figure 8.1 reminds us that the process of engineering design begins with divergent thinking. A problem or a need—for example, "how can I distribute baked beans to consumers?"—arises, for which we desire to create a technological solution. Traditional definitions of divergent thinking—for example, "thinking... that generates a variety of

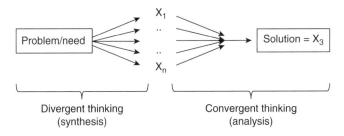

Figure 8.1 Stages in the design process.

ideas" (Russ & Fiorelli, 2010, p. 236) captured in the Torrance Tests of Creative Thinking (Torrance, 1966)—usually illustrate this process in the following way. As part of an "Alternate Uses" test, participants may be invited to think of as many uses as they can for a tin can. While there is no doubt that such a question tests divergent thinking (for example, a tin can may be used as a suit of armor for a mouse, a cup for drinking, or one end of a communication device), divergent thinking in engineering design (and, arguably, in any practical problem-solving context) is manifest in a subtly different way. To illustrate the difference, consider the fact that engineers rarely select an object, for example a tin can, and ask: "What are all the possible things I could do with this object?" Instead, the more typical design process is that a question is asked or a problem posed—for example, "How can I distribute baked beans to consumers?"—and a variety of possible solutions are proposed (such as, in a plastic bag, in my hand, or in a tin can). Both examples represent divergent thinking in the sense that a variety of ideas is generated in response to a question or problem. However, in the former case, the progression is really from solution (tin can) to possible needs, akin to a version of the game show *Jeopardy*, in which there are many correct questions in response to the given answer. By contrast, in the engineering design example, the progression is from need (distribute baked beans) to possible solutions (including, among other things, a tin can). In engineering parlance, we can regard this as the difference between bottom-up design ("What need can I satisfy with this object?") and top-down design ("How can I satisfy this need?").

The difference may seem trivial; and indeed, it is of little consequence to the definition of divergent thinking. However, in the context of the complete design (or creative problem-solving) process, where divergent thinking is followed by convergent thinking, it highlights an element that is critical to any discussion of the interaction between engineering design, creativity, and ethics. Divergent thinking, in isolation, is free to consider *any possible* solution that enters the mind of the designer. Thus, there is, for example, no limit to the *possible* uses of a tin can, and no limit to the number of ways in which baked beans *could be* distributed to consumers. However, divergent thinking, as the precursor to convergent thinking, and taking place within a creative problem-solving (or top-down engineering design) process, does not have the same freedom. While it is true that there may be an infinite number of ways in which baked beans *could be* distributed to consumers, achieving a practical solution dictates that many of these options will be rejected during the stage of convergent thinking.

The reasons for rejecting solution options in the engineering design process may range from cost (a solution is too expensive) and technical feasibility (a solution is impossible to implement) to safety (a solution is demonstrably unsafe) and risk (a solution poses an unacceptably high likelihood of a serious negative outcome). Each of these parameters introduces constraints that, in effect, limit the available range of solutions from all *possible* solutions to a subset of feasible, *practical* solutions. Considerations of safety and risk are, at their core, questions of ethics and morality.

Freedom versus constraint

The preceding discussion again highlights the paradox that is central to this chapter. At the heart of top-down engineering design specifically, and creative problem-solving more generally, is the ability to generate (Guilford, 1950, 1967) many different ideas (fluency), of different types (flexibility), that are unusual (originality), and to develop these ideas (elaboration). This divergent thinking characterizes creativity and depends, for its success, on freedom from constraint. The only way in which we hope to identify and develop effective, competitive, technological solutions to engineering problems is to explore the largest possible design space; that is, to maximize fluency, flexibility, originality, and elaboration. At the same time, in practice, we are bound by constraints that place limits on that design space. Successful design is contingent on maximizing the design space, while practical limitations act to minimize the design space. It would appear, therefore, that it is not possible ever to realize fully successful design, because constraints do not permit the designer to explore the unfettered, maximum *theoretical* design space, in which reside the highly effective and novel solutions that satisfy needs and capture new markets. The designer must, instead,

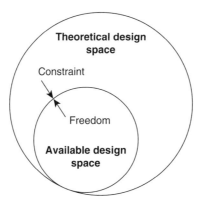

Figure 8.2 The *theoretical* and *available* design spaces.

settle for a limited and unsatisfactory *available* design space that is more likely to be filled with routine solutions that, while probably effective, lack the novelty that opens up new markets and new possibilities (see Figure 8.2).

Constraints and engineering design

The creativity literature rightly devotes a great deal of attention to the range of factors that can inhibit or foster the generation of effective novelty. From early in the modern creativity era, following Guilford's (1950) seminal work, research has examined four main facets of creativity: person, product, process, and press (MacKinnon, 1978; Rhodes, 1961). The constraints that inhibit creativity, therefore, can be grouped under these categories.

Osborn (1953), for example, described the importance of process and sought to address the constraints associated with the misapplication of divergent and convergent thinking by defining a strict process in the form of *brainstorming*. Indeed, Osborn's description of the creative problem-solving process could be characterized as a model of stage-specific constraints based on the 4Ps. More recent versions of the creative problem-solving process include Puccio and Cabra (2009), Puccio, Murdock, and Mance (2005), and Isaksen and Treffinger (2004).

Cropley and Cropley (2012) have described a *phase model* of the process of innovation that illustrates the relationship between the stages of creative problem-solving and the 4Ps (see Figure 8.3). For example, during the stage of generation, they argue that divergent thinking is the dominant process, while in the subsequent stage of illumination, convergent thinking is most active. They reason, therefore, that a stage-specific constraint at this point in the process would be the potential barrier in transitioning from divergent to convergent thinking. In other words, premature analysis, judgment, and criticism (that is, convergent thinking applied too early) would interfere with the successful execution of the generation stage, preventing the exploration of the fullest range of possible solutions during this stage.

Creative problem-solving, in the general sense, seeks to address this conflict between freedom and constraint by quarantining the stages of the process from each other, minimizing any contamination of one by the other. Similarly, it is possible to identify other stage-specific constraints in this model, derived from the relationship between the 4Ps and the stages of innovation (Figure 8.3).

Ethical constraints and engineering design

In addition to the stage-specific constraints described in the previous section, we propose that there are also stage-independent constraints that arise across all stages of the design process, which can be traced to aspects of person, process, press, and product. In engineering design, the impact of morality and ethics is an example of this. To understand the impact of ethics on engineering design, and the role of ethics in constraining the engineering design space, it is necessary to consider four common ethical theories.

Utilitarianism

This theory, originating with the philosopher John Stuart Mill (2007), seeks the maximization of human wellbeing. It is a collectivist approach to ethics in that the focus is on the wellbeing of an entire society and not of individuals. Utilitarianism can be regarded as concerning the

	Preparation	Activation	Generation	Illumination	Verification	Communication	Validation
Process			Divergent thinking	Convergent thinking			
Product							
Person							
Press			Low demand	High demand			

Figure 8.3   Stage-specific constraints on engineering design.

balance between good and bad consequences, or benefits and costs, to society that arise from solutions to engineering problems. Two forms of utilitarianism—act and rule—also have been described (Fleddermann, 2004). Act utilitarianism applies to the outcomes of design (products) and dictates that products should be judged based on whether the greatest good results from a given situation, allowing for rules to be broken to achieve this. Rule utilitarianism, by contrast, dictates that moral rules have primacy. In engineering, a moral rule such as "do not harm people" transforms into a design rule requiring the same, and places a global constraint on the design process. No matter what stage of the design process is active, rule utilitarianism dictates that no engineering solution will be considered that is likely to lead to death or injury. Thus, it permanently closes off certain regions of the design space as *no-go areas*, no matter how creative they might be. Utilitarianism may be considered a source of *stage-independent constraint* on the design process shaped by the product.

Duty ethics

This ethical theory, regarded as originating with the philosopher Immanuel Kant (Kant, Wood, & Schneewind, 2002), can be associated with the person in our stage model of creativity and innovation. In duty ethics, we think of a set of ethical actions that can be written as a list of duties. If the individual (and the organization) follows his or her duties—for example, "be honest," "don't cause harm to others"—then the actions that result, in this case in the design process, will be ethical. Thus, the duty to cause no harm to others leads to both the individual and the organization developing solutions to problems that are safe. In the context of the design process, this represents a set of constraints that derive from the person and the press, which exert influence across the entire design process. Certain solution options are never considered because they would violate the duties that are spelled out for individuals and organizations.

Rights ethics

This ethical theory is closely associated with duty ethics, but is framed from the point of view of the customer in the context of the engineering design process (Fleddermann, 2004). Whereas duty ethics spells out the fact that individuals (and organizations) have a duty to protect others, rights ethics spells out the rights of individuals that others have a duty to protect. The impact is essentially the same for the engineering design process. The duties of the individual designer or design organization, and the rights of consumers of the outcome of design, combine to create a set of constraints that limit what can and cannot be considered in the engineering design space. Even if, by closing off this part of the design space, we lose access to a range of solutions that meet other criteria, we cannot consider them at any point because they would violate fundamental, unbreakable constraints.

Virtue ethics

Virtue ethics (Hursthouse, 1999) holds that "actions are considered right if they support good character traits (virtues), and wrong if they support bad character traits (vices)" (Fleddermann, 2004, p. 38). Responsibility and honesty are important virtues that flow through into engineering design.

Corporate morality and moral agency

Fleddermann raises an important issue that adds a level of complexity to the question of ethics in engineering design and constraints on creativity. In simple terms, "Is there a difference between the ethics practiced by an individual and the ethics practiced by a corporation?" (2004, p. 38). If we are correct, and the press is a source of stage-independent constraints in engineering design (see Figure 8.4), then organizational behavior must play a role, and corporations must have some degree of moral agency. It must be possible to hold a corporation to account for its actions. However, doing so creates further constraints on the available design space in engineering design.

Safety and risk in engineering design

Safety and risk play an especially significant role in constraining engineering design. They represent an outward manifestation of all of the ethical theories. Whether prompted by process (design rules traced to rule utilitarianism), person (good behaviors traced to virtue and duty ethics), press (good corporate citizenship also traced to virtue and duty ethics in the context of moral agents), or product (cost–benefit considerations derived from general utilitarianism), safety and risk stand out as significant constraints on design. For example, if safety and risk were not factors in automobiles, then designers would be free to consider solutions to the underlying need that encompassed much higher speeds, lighter-weight materials, an absence of safety devices such as seat belts, more aerodynamic shapes, more energetic fuels, and so on.

	Preparation	Activation	Generation	Illumination	Verification	Communication	Validation
Process					*Rule utilitarianism* (Moral rules lead to design rules = constraint)		
Product					*General utilitarianism* (Maximizing benefit to society = constraint)		
Person				*Virtue ethics* (Virtuous engineers do not consider options that are irresponsible, dangerous = constraint)			
				Duty ethics (Engineers have a duty to protect the rights of consumers = constraint)			
Press				*Virtue ethics* (Virtuous organizations seek to look after their customers = constraint)			
				Duty ethics (Moral agents seek to protect the rights of consumers = constraint)			

Figure 8.4   Stage-independent constraints and theories of ethics.

Considerations of safety and risk, however, do exert an influence on the design space, ruling out many of these options. While there is an implied warranty in engineering design (namely, that products will be safe to use), a dilemma faced by designers is that nothing can be completely safe. Engineers, however, are required to make products as safe as reasonably possible. For this reason, safety plays a pervasive, constraining role in engineering design. To judge whether a design is safe or not, Fleddermann (2004) describes four criteria that must be met. We can see links to the Kansas City Hyatt walkway collapse in each of these. The first criterion is that, as a minimum, a design must comply with applicable laws. The second criterion is that a design should meet accepted standards. The third criterion is that "alternative designs that are potentially safer must be explored" (2004, p. 65). This particular criterion is significant because the exploration introduces an opportunity for creativity, and it is this reintroduction of creativity that gives us an opportunity to expand the available design space back into the theoretical design space (see Figure 8.2). The fourth criterion is that engineers must attempt to predict the ways in which a product might be misused by the consumer and then design a product to avoid these problems. This criterion also introduces an opportunity for creativity that rebalances the available design space.

The reaction to these latter two criteria is perhaps the most obvious way in which engineers can offset the effects of ethical constraints on the theoretical design space. Although the ethical constraints initially appear to limit the design space to a subset of the theoretical maximum, the introduction of new design problems based on these criteria for safety in fact represents a reformulation of the design problem. We will revisit this issue shortly.

In the language of system design and requirements, the first step in design is identifying a set of requirements—a set of statements that expresses the purpose of the thing to be created and constraints on that design. In engineering design, there are at least two ways in which ethics plays a role here. First, bearing in mind that requirements form a set of criteria by which a product is tested, ethics drives a move to ensure that a full set of requirements even exists. In other words, a narrow view would be to say that because there was no explicit, expressed requirement for the system to be safe, there was no test in relation to safety that was failed. Therefore, the system meets all the requirements and is satisfactory. This is a sin of omission. Just because there is no requirement for safety, for example, does not mean that the system is inherently safe. An ethical framework, therefore, demands that the set of requirements that dictates the design space is, by definition, complete in all respects,

including covering things like safety. Second, there is the specific introduction of ethically driven requirements that limit the design space.
In the language of engineering design, we talk of non-functional system
(NFS) requirements, those specific statements that spell out constraints
on the whole system. These include safety, cost, size, weight, and a range
of other factors often call "ilities," such as maintainability, reliability,
and so on. Not only do these NFS requirements impose constraints, or
limitations, on the design space, shrinking it to a subset of the possible
solutions, they also introduce the possibility of trade-offs. The implication of trade-offs is that certain qualities of the solution—its cost, size,
weight, and safety—are negotiable and, indeed, frequently in conflict.
Take as an example the famous case of the Ford Pinto. Fleddermann
(2004) points out that there were, in effect, two competing NFS requirements at odds: cost and safety. Setting an upper limit on cost imposed a
practical upper limit on safety that fell below the minimum required for
actual safe operation. The only way in which designers could balance
this trade-off was to violate one of the NFS requirements (in this case,
safety). The upper limit on cost directly constrained the design space to
a subset of solution options that were unsafe.

This situation creates an apparent paradox. The only way to satisfy the
requirement for safety (apart from ignoring it) is to violate the requirement for cost. In the case of the Ford Pinto, management dictated that
cost was the primary consideration, so the engineers were faced with
the dilemma of how to satisfy the requirement for safety. The solution,
in plain terms, was to soften the requirement for safety to the point
that the chosen solution met it. The role of creativity here is to reex-
pand the design space (shrunk by the cost and safety constraints) back
to the point at which it included safe solutions. However, such a requirement limits the design space to a subset of the full, potential design
space—in other words, only those solutions that meet the ethically
based requirements such as safety.

Tackling ethical constraints in engineering design

Engineering design problems begin with a theoretically unconstrained
design space, where the opportunity for creativity is at a maximum.
This is then reduced to an available design space (Figure 8.2) by a
range of ethical and other constraints. Because of a desire to recognize and realize the benefits of creativity, engineers nevertheless seek
to offset the effect of those constraints. This can be done in one of
two ways. Either the engineer can ignore the ethical constraints (both
statutory and customary)—and there are examples from the history of

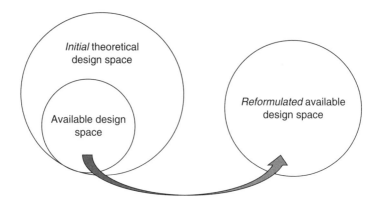

Figure 8.5   Sidestepping constraints—reformulating the problem.

engineering design where this pathway has been followed, although fortunately these are relatively rare—or the engineer can rebalance the ethical constraints by shifting the design problem in new directions. Thus, what appears at first sight to be a constraint—for example, safety—in fact can be seen to function more as a stimulus for redefinition of the problem, for which there is a new and different available design space that is less constrained than the original design space. This sidestep can be illustrated as in Figure 8.5.

Therefore, freedom and constraint in the context of engineering design are, in fact, entirely reconcilable. Constraints, perhaps, need to be seen as highly contextual in nature. A constraint is only a constraint under a certain set of conditions. It may be possible to remove, or at least sidestep, a constraint in a manner that does not violate overarching ethical considerations. It is entirely possible to devise ethical *and* creative engineering solutions. The limitations imposed by ethical frameworks do not constrain design to the point at which no creativity is possible. When engineers reframe the problem, it is still possible to generate effective novelty. The requirement for ethics, although it will always limit to some degree the range of solution options, does not militate against creativity, but it can still serve as a stimulus for creativity, provided we understand how the process operates.

Conclusion

The complex interrelationship of novelty, effectiveness, and technical engineering problem-solving creates an apparent paradox. That paradox is expressed in the tension between a desire for freedom and a requirement for constraint. In the engineering profession, ethics describes the set of rules and standards that governs the conduct of engineers in their capacity as professionals. A framework of ethics and morals, in particular, addresses the potential for harm that exists when engineering solutions are created, and limits what can be done. At the same time, the defining creative and value-adding activity of engineering—design—demands patterns of thought and behavior, both in the individual and the organization, that favor risk-taking, freedom from constraint, and tolerance of uncertainty. How is it possible to balance the need for freedom that creative design requires with the inherent constraints imposed by ethics?

The solution to this paradox may lie in how we make sense of those constraints. It is tempting to see constraints only in terms of how they close off and limit the design space (Figure 8.2). This view, however, ignores the creativity that can also be applied to how each problem is formulated. If constraints, whether arising from ethical considerations or not, are first seen as a stimulus to reformulating the problem and not simply a barrier to design, then the design space is *not* reduced. Instead, the design space shifts, or sidesteps, into a region that is unconstrained or less constrained compared to the original (Figure 8.5).

The necessity that gives rise to invention is not so much in the original need as in the flexibility and creativity that engineers apply as they reformulate a problem to accommodate what can and cannot be done. Once we see constraints as a stimulus to creativity and not an impediment, it becomes clear that it *is* possible to devise ethical, creative engineering solutions, and that the requirement for ethics does not militate against creativity.

References

Blumrich, J. F. (1970). Design. *Science, 168*, 1551–1554.

Bruner, J. S. (1962). The creative surprise. In H. E. Gruber, G. Terrell, & M. Wertheimer (eds.), *Contemporary Approaches to Creative Thinking* (pp. 1–30). New York: Atherton.

Cropley, A. J. (1990). Creativity and mental health in everyday life. *Creativity Research Journal, 3*(3), 167–178.

Cropley, A. J. (2010a). The dark side of creativity. In D. H. Cropley, A. J. Cropley, J. C. Kaufman, & M. A. Runco (eds.), *The Dark Side of Creativity* (pp. 1–14). New York: Cambridge University Press.

Cropley, D. H. (2010b). Summary—The dark side of creativity: A differentiated model. In D. H. Cropley, A. J. Cropley, J. C. Kaufman, & M. A. Runco (eds.), *The Dark Side of Creativity* (pp. 360–374). New York: Cambridge University Press.

168   *The Ethics of Creativity*

Cropley, D. H., & Cropley, A. J. (2005). Engineering creativity: A systems concept of functional creativity. In J. C. Kaufman & J. Baer (eds.), *Faces of the Muse: How People Think, Work and Act Creatively in Diverse Domains* (pp. 169–185). Hillsdale, NJ: Lawrence Erlbaum Associates.

Cropley, D. H., & Cropley, A. J. (2012). A psychological taxonomy of organizational innovation: Resolving the paradoxes. *Creativity Research Journal, 24*(1), 29–40.

Cropley, D. H., & Cropley, A. J. (2013). *Creativity and Crime: A Psychological Approach.* Cambridge: Cambridge University Press.

Cropley, D. H., Kaufman, J. C., & Cropley, A. J. (2008). Malevolent creativity: A functional model of creativity in terrorism and crime. *Creativity Research Journal, 20*(2), 105–115.

Cropley, D. H., Cropley, A. J., Kaufman, J. C., & Runco, M. A. (eds.). (2010). *The Dark Side of Creativity.* New York: Cambridge University Press.

Dieter, G. E., & Schmidt, L. C. (2012). *Engineering Design,* 5th edn. New York: McGraw-Hill Higher Education.

Fleddermann, C. B. (2004). *Engineering Ethics,* 2nd edn. Upper Saddle River, NJ: Pearson Prentice Hall.

Guilford, J. P. (1950). Creativity. *American Psychologist, 5,* 444–454.

Guilford, J. P. (1967). *The Nature of Human Intelligence.* New York: McGraw-Hill.

Horenstein, M. N. (2002). *Design Concepts for Engineers,* 2nd edn. Upper Saddle River, NJ: Prentice-Hall.

Hursthouse, R. (1999). *On Virtue Ethics.* Oxford: Oxford University Press.

Isaksen, S. G., & Treffinger, D. J. (2004). Celebrating 50 years of reflective practice: Versions of creative problem solving. *Journal of Creative Behavior, 38*(2), 75–101.

Kant, I., Wood, A. W., & Schneewind, J. B. (2002). *Groundwork for the Metaphysics of Morals.* New Haven, CT: Yale University Press.

MacKinnon, D. W. (1978). *In Search of Human Effectiveness: Identifying and Developing Creativity.* Buffalo, NY: Creative Education Foundation.

Martindale, C., & Moore, K. (1988). Priming, prototypicality, and preference. *Journal of Experimental Psychology: Human Perception and Performance, 14*(4), 661–670.

Mill, J. S. (2007). *Utilitarianism, Liberty and Representative Government.* Rockville, MD: Wildside Press.

Oral, G. (2006). Creativity of Turkish prospective teachers. *Creativity Research Journal, 18*(1), 65–73.

Osborn, A. F. (1953). *Applied Imagination.* New York: Scribners.

Plucker, J. A., Beghetto, R. A., & Dow, G. T. (2004). Why isn't creativity more important to educational psychologists? Potentials, pitfalls, and future directions in creativity research. *Educational Psychologist, 39*(2), 83–96.

Puccio, G., & Cabra, J. (2009). Creative problem solving: Past, present and future. *The Routledge Companion to Creativity* (pp. 327–337). Oxford: Routledge.

Puccio, G. J., Murdock, M. C., & Mance, M. (2005). Current developments in creative problem solving for organizations: A focus on thinking skills and styles. *Korean Journal of Thinking and Problem Solving, 15*(2), 43–76.

Rhodes, M. (1961). An analysis of creativity. *The Phi Delta Kappan, 42*(7), 305–310.

Russ, S. W., & Fiorelli, J. A. (2010). Developmental approaches to creativity. In J. C. Kaufman & R. J. Sternberg (eds.), *The Cambridge Handbook of Creativity* (pp. 233–249). New York: Cambridge University Press.

Salcedo-Albarán, E., Kuszewski, A., de Leon-Beltran, I., & Garay, L. J. (2009). Rule-breaking from illegality: A trans-disciplinary inquiry. METODO Working Papers No. 63. Bogotá: Grupo Metodo.

Silvia, P. J., & Brown, E. M. (2007). Anger, disgust, and the negative aesthetic emotions: Expanding an appraisal model of aesthetic experience. *Psychology of Aesthetics, Creativity, and the Arts, 1*(2), 100–106.

Torrance, E. P. (1966). *Torrance Tests of Creative Thinking: Technical Norms Manual.* Lexington, MA: Personnel Press.

9

Construction or Demolition: Does Problem Construction Influence the Ethicality of Creativity?

Daniel J. Harris, Roni Reiter-Palmon, and Gina Scott Ligon
University of Nebraska Omaha, USA

Creativity researchers have long sought to understand the factors that foster or inhibit creative performance. Researchers have evaluated individual difference variables, such as personality (Feist, 1998; Furnham & Nederstrom, 2010), motivation (Amabile, 1996), and intelligence (Furnham & Nederstrom, 2010), as well as environmental factors, such as leadership (Reiter-Palmon & Illies, 2004) and organizational culture (Hunter, Bedell, & Mumford, 2007). More recently, researchers also have been interested in the implications or intent of the creative idea or solution. Do solutions lead to negative consequences that can be anticipated or even wanted by the problem-solver? Can we find a way to develop ethical and creative solutions?

Researchers have also studied the role of cognitive processes in creative problem-solving over the last few decades, and this cognitive approach to creativity suggests that one important factor in facilitating creativity is the effective application of a number of cognitive processes (Mumford et al., 1991). The cognitive process approach suggests that creativity—an idea or solution that is of high quality and high originality—results when these cognitive processes are used effectively by problem-solvers. Much of the research in this area has focused on what factors may facilitate or inhibit the effective application of these processes, as well as isolating and determining the effect of each process on creative problem-solving (Mumford et al., 1997; Reiter-Palmon et al., 1997). In this chapter, we focus on the cognitive approach to creativity, specifically on one process: problem construction. We relate this process to ethical decision-making and, finally, suggest directions for future research in which we will explore the role of problem construction in ethical and malevolent creativity.

Problem construction

There are a number of models that focus on the role of cognitive processes in creative problem-solving (Basadur, 1997; Mumford et al., 1991; Ward, Smith, & Finke, 1999). One commonality across these models is that they all suggest that creative thought includes an early stage of problem identification, problem finding, problem definition, problem formulation, or problem construction. Although different models and approaches use different terms, to prevent confusion we will use the term "problem construction" here. Problem construction occurs when problems are ill defined and, therefore, do not have an obvious goal or can have multiple goals. During this process the problem to be solved is recognized and identified (is there indeed a problem?), defined (what is the nature of the problem?), and constructed (what are the parameters of the problem to guide possible solutions?).

Research on problem construction indicates that creative individuals engage in the process more deliberately and more effectively than individuals who are less creative (Getzels & Csikszentmihalyi, 1976; Rostan, 1994). In their seminal study, Getzels and Csikszentmihalyi (1976) evaluated art students' problem construction by observing their behavior prior to painting. Creativity was judged by art experts evaluating the painting, and the results indicated that more time and more effort spent in constructing the problem were related to the aesthetic quality and originality of the resulting painting. Moreover, the problem-construction measure was also related to both financial success and critical success as an artist a number of years later. Other research suggests that experts engage in the problem-construction process in a more deliberate and effortful manner than novices (Voss et al., 1991). These studies indicate that both experts and creative individuals engage in more deliberate problem construction, resulting in more effective and creative products or solutions.

Other research has focused on the link between creativity and problem construction as a stable characteristic. Artley et al. (1980) found that fluency in problem-finding was correlated with fluency in verbal divergent thinking tests in a sample of college students. Okuda, Runco, and Berger (1991) evaluated problem-finding, creative problem-solving for real-world problems, and standard divergent-thinking tests as predictors of participation in creative activities in children. They found that problem-finding was the best predictor of creative accomplishments and added significantly above and beyond the real-world creative problem-solving and divergent-thinking tasks. Reiter-Palmon and colleagues (Reiter-Palmon et al., 1997; Reiter-Palmon, Mumford, & Threlfall, 1998) found that problem-construction ability was correlated with both quality and originality of solutions generated in response to a variety of real-world problems. Mumford et al. (1997) also found that problem-construction ability was predictive of solution quality and originality across two different kinds of real-world problems. More importantly, problem construction was an important predictor even when other creative processes (that is, idea generation) and basic abilities (that is, intelligence) were taken into account.

In addition, the importance of problem construction for creativity was established by evaluating the effect of inducing active engagement in the process. Active engagement has typically been induced by asking participants to restate the problem in multiple ways prior to solving the problem (Baer, 1988) or by asking participants to generate goals and/or constraints for solving the problem (Wigert & Reiter-Palmon, 2013). Research on problem construction suggests that inducing active engagement in the process results in increased creativity compared to those participants who were not instructed to do so (Mumford, Reiter-Palmon, & Redmond, 1994; Reiter-Palmon et al., 1997; Wigert & Reiter-Palmon, 2013). Likewise, the quality and originality of the problem restatements generated for a specific real-world problem were directly related to the quality and originality of the solution generated to the same real-world problem, even after problem-construction ability was taken into account (Arreola & Reiter-Palmon, 2013).

The importance of active engagement is further supported by research on the effects of training in problem construction on creativity. Research on training problem construction, not surprisingly, finds an effect for this specific process (Fontenot, 1993; Kay, 1991). More importantly, training problem construction was related to improved overall creative problem-solving (Basadur, Graen, & Green, 1982; Ellspermann, Evans, & Basadur, 2007). Finally, a meta-analysis on creativity training and its relation to creativity outcomes found that a focus on problem construction was related to improved training outcomes and creative performance across multiple studies (Scott, Leritz, & Mumford, 2004). Although the review of the literature indicates the importance of problem construction for creativity, limited research has focused on the intersection between problem construction and either malevolent creativity or ethical problem-solving. For example, Illies and Reiter-Palmon (2008) found that when problems were constructed in ways that reflected self-enhancement, solutions generated to organizational problems were more destructive. This study indicates that the manner in which problems are constructed is important for understanding the ethicality of the solution.

Mumford, Reiter-Palmon, and Redmond (1994) developed a model of how problems are constructed. They suggest that problem construction is based on problem representations developed from past experiences. Specifically, when individuals solve a problem, they create a cognitive structure that reflects the problem-solving effort (Holyoak, 1984). Problem representations include information about problem-solving goals, information and procedures used, and any restrictions and constraints placed on the solution. Mumford, Reiter-Palmon, and Redmond's (1994) model further suggests that problem representations will be elicited based on cues in the environment related to the problem. This model indicates that individuals will use past problem-solving experiences to guide current problem-solving efforts. When encountered problems are similar to problems that the individual has satisfactorily solved in the past, the problem representation for that particular problem-solving effort will be elicited and used. However, when a problem is novel and ambiguous, multiple problem representations may be activated and incorporated into a new way of constructing the problem. When the problem is ambiguous, and therefore the cues are ambiguous, similarity to past problem-solving efforts and past experiences will guide both the interpretation of the cues and the problem representations that are elicited. One of the important implications of the model suggested by Mumford, Reiter-Palmon, and Redmond (1994) is that past experiences will have considerable impact on the interpretation of cues that elicit problem representations, as well as the actual problem representations that are elicited.

Past experiences

When faced with ill-defined events, individuals rely on a mental model, or cognitive representational system, to define the nature of the problem at hand (Holyoak & Thagard, 1997; Johnson-Laird, 1983). These mental models provide conceptual maps of interrelationships among prior goals and actions in a person's experience in a way that is used to understand a new problem and to guide responses to it (Sein & Bostrom, 1989). For example, if an individual has successfully solved numerous past problems with the use of aggression, then he or she will likely continue

to be guided by that aggression when attempting to solve new problems. Thus, similarities between current ill-defined problems and past
problems may trigger the identification of some or all of an individual's
causal system elements, which are then activated and provide a vehicle
for goal attainment. Mental models serve as a heuristic device for goal
attainment to understand a new problem by comparing it to elements
of previous experiences (Keller, 2003).

Although it holds that past experiences shape how individuals construe the problems they face, it also follows that some experiences are
more impactful than others. Moreover, experiences that have recurring
themes and meanings have more valence in shaping how individuals
use them in defining the current problems they encounter (Bluck &
Habermas, 2000). These thematically related events, or life narratives
(Anderson & Conway, 1993), are not purely cognitive in nature but
instead are combinations of cognition and emotion that can be used
to understand and make sense of problems that are faced later in life
(Pillemer, 2001). McAdams' (2001) research demonstrates that these life
narratives allow for the formation of a personal identity, which provides
a framework for acceptable parameters of performance. For example,
Indira Gandhi (Prime Minister of India from 1966–77 and 1980–84)
had recurring thematic undertones in her life history where she saw
important people around her (for example, her father) recover from
hardship and negative events. Thus, she constructed a personal identity
reflective of redemption and endurance. When faced with challenges in
subsequent years as an adult, she often drew from these life lessons to
remind herself that she was to persevere. Such life narratives serve direc-
tive functions, providing lessons in an episodic, parable-like form that
depicts life lessons used to define goals, causes, actions, and context in
present situations (Baumeister & Newman, 1994).

Given the impact of past experiences on understanding current problems, a great deal of research has examined the content, structure, and
development of the life narrative as it relates to differential decision-
making. Bluck and Habermas (2000) identified four coherent mechanisms that are used to organize past experiences into a life narrative for
subsequent decision-making. First, life narratives are contextually linked
to cultural expectations about the acceptability of a given experience.
For example, experiences with martyrdom are culturally distinct across
regions and accordingly viewed differentially positively or negatively;
thus, themes about such events are anchored to the cultural context in
which they were experienced. Second, life narratives are often related
to a temporal imposition of other important life tasks (for example,

graduation, marriage). Third, narratives are often based on the abstraction of general causes of important life events, particularly steps that are perceived as goal relevant or consequential. For example, events leading up to the decision to embark on a particular career are often later reflected on as important determinants in shaping an individual (for example, the presence of a particular mentor, losing a job prior to finding a new one). Fourth, life narratives are often demonstrative of themes, or recurring life lessons, that bind causes, outcomes, and events together. In McAdams' (2001) work, he finds that similar life lessons are repeated throughout a life narrative, implying some levels of coherence across seemingly discrete events (for example, a series of unrelated events teaching the same lesson).

Bluck's (2003) theory implies that coherent life narratives often arise in late childhood and early adulthood, when abstract reasoning, or the capacity to draw inferences about relationships between events and consequences, fully comes online in individuals. Simonton's (1994) work on creativity also supports the notion that events that occur in adolescence are statistically related to instances of greatness. For example, in his study of European monarchs, Simonton (1983) found that experiences of early role modeling and exposure to high-achieving peers were related to creative and influential performance in leadership roles. In addition, in examining career experiences of eminent scientists and other creative professionals, Mumford et al. (2005) found that significant events (for example, high-performing mentors) in early career were related to subsequent creative problem-solving throughout the lifespan. What all of this research indicates is that one's life history can have a very strong impact on the ways in which future problems will be constructed and responded to.

Leaders' experiences and ethics

One avenue for investigating life experiences and narrative memory, as it relates to ethical decision-making, is to examine historically notable and destructive individuals who have a great deal of objectively verifiable information compiled about their early experiences and the ethicality of subsequent decisions they made (Ligon, Harris, & Hunter, 2012). While it is difficult to generalize findings from such extreme examples to everyday problem construction, the study of historically notable figures, such as destructive leaders, provides an empirical avenue to identify some relationships between life experiences and subsequent problem construction and decision-making. Krasikova, Green, and LeBreton define destructive leaders as individuals who "a) encourage followers to pursue destructive goals (for example, Hitler), and/or b) use destructive methods of influence with followers (for example, leaders who bully followers)" (2013, p. 5). Examining the life narratives and past experiences of such destructive leaders can provide some evidence about the mental model formation they might have used in problem construction.

For example, in a seminal study that investigated differences between ethical and unethical charismatic leaders, O'Connor et al. (1995) examined the life histories of individuals who were more interested in their own personal outcomes (that is, unethical or personalized leaders) versus group outcomes (that is, ethical or socialized leaders). The authors found that leaders who had early experiences with negative life themes, object beliefs, and personalized need for power were more likely to make impetuous decisions with destructive consequences (House & Howell, 1992; McClelland, 1975). While ethical leaders used their position of power to serve others, unethical leaders used their position of power for personal gain and self-promotion, construing situations as opportunities for other-exploitation and self-aggrandizement (Howell & Avolio, 1992). That situational construal can be likened to problem construction: unethical leaders tend to construct problems in ways that facilitate the exploitation of others (for example, followers) and gain more power for themselves. O'Connor et al. (1995) also found evidence for early experiences of powerlessness. Related clinical work suggests that enduring early experiences with prolonged powerlessness are associated with later decisions as an adult to use coercive influence techniques (Goodstadt & Hjelle, 1973), ignore the feelings of others, and exploit people for personal gain (McClelland, 1975; Rosenthal & Pittinsky, 2006).

In a review of 120 historically notable leaders, Ligon, Hunter, and Mumford (2008) found that experiences with recurring themes of contamination, or seemingly positive experiences that turned out to have negative consequences (McAdams, 1998), were related to destructive decision-making in later life. Specifically, certain events were "surprising and negative" in that they seemed innocuous but turned out to have deleterious consequences. Individuals who experienced several of those types of life events went on to view people (and situations) as things to be managed and conquered. For example, Fidel Castro, after experiencing continued humiliating events at school, later reflected back on those life lessons to infer that he would always dominate those who surrounded him in order to prevent the potential for later mistreatment from them. Such contaminating events may be particularly important in problem-construction activities as they generalize the need to defend against unanticipated, harmful, and humiliating consequences before they occur. Thus, problems are construed in a way that is aggressive, defensive, and skeptical. This *a priori* defensive posture likely permits the expression of malevolent or unethical problem-solving, as well as permitting others to do so. One way of explaining such destructive problem construction is through the use of justification mechanisms.

Justification mechanisms

Justification mechanisms guide behavior to the extent that they allow certain behavioral tendencies to be regarded as rational and logical (James, 1998). Justification mechanisms are named as such because they are unconscious cognitive mechanisms that allow people to justify their behaviors, such as justifying why a prosocial behavior (for example, altruism) is appropriate in a given situation versus justifying why an aggressive behavior (for example, mockery) is appropriate. Because certain behaviors are viewed as rational via the use of justification mechanisms, those particular behaviors are solidified as the preferred ways in which some people interpret and react to the world around them. Using different justification mechanisms (for example, justification mechanisms for prosocial vs. aggressive behavior) is called conditional reasoning, which suggests that people differ with regard to which behaviors they deem justifiable in certain contexts (James, 1998).

Although aggressive behavior is often not appropriate in a number of social situations, aggressive individuals likely justify their aggression not only as being logical in general, but in some instances as being the *most* logical course of action that could be taken. Some examples of justification mechanisms for aggression include an increased likelihood of seeing malevolent intent in others' actions, even in good or well-meaning actions; having a predisposition to retaliate rather than reconcile after a perceived offense; and framing oneself as a victim and framing others along a spectrum of weakness and strength (James, 1998). Aggressive people probably interpret social situations and interactions so differently from prosocial people that aggressive people might view prosocial justification mechanisms as being highly illogical and irrational. Aggressive people probably do not rely solely on aggressive justification mechanisms to guide their behavior; instead, it is likely that individuals who more strongly justify their aggressive acts have stronger dispositions to be aggressive (James, 1998).

Because justification mechanisms allow people to determine what they believe to be appropriate courses of action, justification mechanisms likely influence the ways in which problems are constructed. That is, justification mechanisms are biases that guide how individuals construe problems and define parameters for acceptable solutions. Similar to problem representations, justification mechanisms tend to be derived from specific types of past, emotionally evocative experiences. For example, if an employee believes that her co-workers are always trying to exploit her, regardless of the veracity of her perceptions, then she will probably construct work-related problems (for example, interpersonal conflicts) as events that permit, or even require, aggressive behaviors.

With regard to leadership, James and LeBreton (2011) discussed how one's predisposition toward implicitly justifying aggressive behaviors results in destructive leadership, interpreting problems and others' actions as intentionally harmful. Moreover, individuals who impute hostile intent into the behaviors of others are also more likely to interpret everyday situations as ones where their needs are blocked, and tend to feel justified or even proactively motivated to make decisions that would harm others. In essence, individuals who engage in destructive acts probably construe day-to-day interactions with others through cognitive biases (for example, victimization bias, hostile attribution bias) that present a negative view of how the world works and their place in it. Such biases develop over time and through repeated experiences that reaffirm negative life themes and distortions about the motivations of others, and they provide a rationale and justification for *a priori* inflicting harm on others through unethical decisions. One particularly potent way of inflicting harm on others is to be destructive in original ways, which is referred to as malevolent creativity.

Framing and constructing malevolent creativity

Malevolent creativity is defined as "the interaction among aptitude, process, and environment by which an individual or group produces novel and useful ideas as defined within a social context...that are intended to materially, mentally, or physically harm oneself or others" (Harris, Reiter-Palmon, & Kaufman, 2013, p. 237). Harmful acts, such as terrorism, deception, abuse, aggressive humor, and theft, can be malevolently creative (Cropley, Kaufman, & Cropley, 2008), but they must be *novel* instances of such acts. An idea or product is not malevolently creative merely because it is harmful. It is of note that a similar construct, negative creativity, was coined earlier by James, Clark, and Cropanzano (1999). The difference between malevolent and negative creativity is that the former pertains to acts of creativity that are *intended* to harm in some way, whereas the latter pertains to acts of creativity that result in harm yet may not have been intended as such. Beyond those semantic differences, the empirical study of malevolent creativity is very much in its infancy. What is currently suggested by the literature is that perceptions of unfair situations facilitate the generation of negatively (malevolently) creative ideas (Clark & James, 1999); trait physical aggression positively relates to, and conscientiousness negatively relates to, malevolent creativity (Lee & Dow, 2011); and emotional intelligence negatively relates to malevolent creativity even when controlling for cognitive intelligence and task effects (Harris, Reiter-Palmon, & Kaufman, 2013).

Although the relationship between destructive problem construction and malevolent creativity is highly speculative, it can be assumed that people who are more malevolently creative construct problems differently than people who are less malevolently creative. For example, as noted above, people who are more physically aggressive are more malevolently creative (Lee & Dow, 2011), which indicates that there might be a relationship between *overt* aggression and malevolent creativity. However, the use of aggressive justification mechanisms, otherwise known as *implicit* aggression (Bing et al., 2007), may relate to malevolent creativity in different ways. Because implicitly aggressive individuals often frame and structure problems in highly aggressive ways, they likely develop a type of "aggressive expertise" that allows them to think more flexibly and uniquely about aggressive behaviors. De Dreu and Nijstad (2008), for example, found that individuals with conflict-related cognitions generated more original competitive tactics than did individuals with cooperative cognitions. Coupled with the notion that experts engage in more deliberate and effortful problem construction than novices (Voss et al., 1991), it is quite possible that implicitly aggressive individuals are more malevolently creative, in general, because of enhanced problem-construction efforts, and, furthermore, are more malevolently creative in negatively valenced situations that facilitate and promote their implicit aggression.

Harris and Reiter-Palmon (2013) studied the effects of task and implicit aggression on malevolent creativity, and found that task and implicit aggression interacted to influence the generation of malevolently creative ideas. Participants responded to one of two problem-solving tasks: an aggressive problem, which was framed to permit malevolent creativity and tap into highly reactive aggression, or a prosocial problem, which was framed to promote benevolent creativity and tap into a highly cooperative mindset. Prosocial individuals (that is, individuals who do not endorse aggressive justification mechanisms) generated very few, if any, malevolently creative ideas across both tasks.

This result is not surprising because prosocial individuals likely do not view harmful responses as appropriate ways of responding to most socially oriented situations. Similarly, prosocial individuals likely do not have expertise in thinking aggressively, which may constrain the originality of any harmful ideas they do generate. In that way, prosocial individuals probably engage in benevolent problem construction, thinking of tactics to resolve an issue in more ethical, cooperative, or legal ways. Implicitly aggressive individuals, on the other hand, generated more malevolently creative ideas than did prosocial individuals, especially in response to the aggressive problem.

This result also is not surprising because implicitly aggressive individuals probably engage in destructive problem construction, considering and generating ideas that are more harmful, destructive, illegal, and unethical. However, the most fascinating aspect of the interaction between task and implicit aggression is that implicitly aggressive individuals generated very few, if any, malevolently creative ideas in response to the prosocial task. What this suggests is that implicitly aggressive individuals may engage in destructive problem construction only when there are sufficient cues to activate such aggression.

The implications of these different research streams suggest that both ethical and malevolent creativity tend to depend on the kinds of justification mechanisms that are used and how problems are constructed. When past experience leads to the development of justification mechanisms that are prosocial and ethical in nature, individuals will be more likely to construe ambiguous situations in a manner consistent with these justification mechanisms. Similarly, past experiences and life themes of negative events and destructiveness would lead to justification mechanisms and situational construals that promote aggression in a consistent manner. These aggressive or prosocial problem constructions probably indicate an appropriate course of action, or solution, to the individual. Thus, we expect that malevolently creative individuals view the world and construct problems in highly aggressive ways, whereas individuals who develop ethical creative solutions have very different problem constructions—ones that indicate a prosocial understanding and interpretation of a problem.

Future directions and conclusion

Although the literature examining historically notable individuals (for example, leaders) yields some insight into the relationship between problem construction, ethics, and creativity, more empirical work is needed. The first step in determining whether problem construction influences the ethicality of creativity is to study the malevolence of problem restatements (Baer, 1988) or problem goals and constraints (Wigert & Reiter-Palmon, 2013) and whether the malevolence of the problem-construction phase relates to the ways in which individuals respond to problems. Put differently, a broad research question is whether the malevolence found in problem constructions influences the kinds of ideas that are generated, and eventually implemented, in response to problems. Illies and Reiter-Palmon (2008) start to address this issue, and the findings are promising. If destructive problem construction is found to relate to the malevolent creativity of subsequent ideas or products, then mediating and boundary factors, such as implicit and explicit aggression, impulsivity, or emotional intelligence, should be explored. Likewise, certain personality variables that are more negatively valenced (for example, the Dark Triad, which consists of Machiavellianism, narcissism, and psychopathy) should also be examined in relation to how problems are constructed and responded to creatively.

After getting results from laboratory settings (such as paper-and-pencil creative problem-solving simulations), another avenue of research to consider would be real-life instances of destructive problem construction. Certain populations may be more relevant to the study of destructive problem construction than others, such as gang members, prisoners, or terrorists. However, given the difficulties of accessing samples from those populations, another potential area to consider would be historiometric analysis. Historiometry is the quantification of qualitative, historically oriented materials (Ligon, Harris, & Hunter, 2012), such as applying coding schemes to content found in biographies, websites, or court hearings. A study similar to that of Ligon, Hunter, and Mumford (2008) could explore notable leaders via their justification mechanisms and the subsequent malevolent creativity of their followers, or even the malevolent creativity of the leaders themselves. For example, leaders' and followers' actions (described in biographies, websites, and so on) could be rated on their originality and harmfulness to assess malevolent creativity, and communications from leaders could be coded for aggressive justification mechanisms. Figure 9.1 shows an

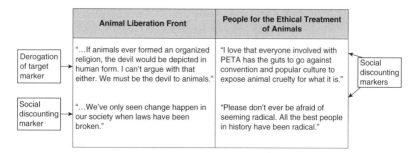

Figure 9.1   Comparison of destructive problem constructions and markers for aggression between leaders of ideological organizations.

example comparison of destructive problem constructions and markers of aggression made by a spokesperson of the Animal Liberation Front and the president of People for the Ethical Treatment of Animals, both in videos posted on YouTube. Having that kind of information could offer insights into specific aggressive justification mechanisms that leaders of violent (vs. non-violent) groups use to justify harming others, especially in creative ways. Furthermore, knowing which types of life-history events best predict the development of aggressive justification mechanisms and malevolent creativity could give researchers better insight into the complex relationship that likely exists between destructive problem construction and malevolent creativity. In particular, security and intelligence agencies would have a better indication as to which extremist leaders might be more willing to engage in violence.

Ultimately, it seems that individuals who are exposed to numerous negative life events may develop more aggressive justification mechanisms to cope with their harsh realities. Such aggressive justification mechanisms could fuel their use of destructive problem construction when encountering future problems. Those problems might be responded to in malevolently creative ways, but the novelty of aggressive behaviors likely depends on the individual's "aggressive expertise" (via implicit aggression) and the degree to which he or she engages in thoughtful, effortful problem construction. Likewise, if those types of individuals become leaders, they may further use aggressive justification mechanisms to exploit others, increase their power base, demean non-members, or even promote and facilitate the use of malevolent creativity against specific targets via the use of destructive problem construction. The cognitive and developmental antecedents of malevolent creativity require further empirical investigation, especially when considering the amount of psychological and physical harm that may result from the actions of those who construct their realities in highly destructive ways.

Acknowledgement

This material is based upon work supported by the U.S. Department of Homeland Security under Grant Award Number 2010-ST-061-RE0001. The views and conclusions contained in this document are those of the authors and should not be interpreted as necessarily representing the official policies, either expressed or implied, of the U.S. Department of Homeland Security.

References

Amabile, T. M. (1996). *Creativity in Context*. Boulder, CO: Westview Press.
Anderson, S. J., & Conway, M. S. (1993). Investigating the structure of autobiographical memories. *Journal of Experimental Psychology: Learning, Memory, and Cognition, 19*, 1178–1196.
Arreola, N. J., & Reiter-Palmon, R. (2013). The influence of personality on problem construction and creative problem solving. Unpublished manuscript.
Artley, N. L., Van Horn, R., Friedrich, D. D., & Carroll, J. L. (1980). The relationship between problem finding, creativity, and cognitive style. *Creative Child and Adult Quarterly, 5*, 20–26.
Baer, J. M. (1988). Long-term effects of creativity training with middle school students. *Journal of Early Adolescence, 8*, 183–193.
Basadur, M. S. (1997). Organizational development interventions for enhancing creativity in the workplace. *Journal of Creative Behavior, 31*, 59–72.
Basadur, M. S., Graen, G. B., & Green, S. G. (1982). Training in creative problem solving: Effects on ideation and problem finding in an applied research organization. *Organizational Behavior and Human Performance, 30*, 41–70.
Baumeister, R. F., & Newman, L. S. (1994). How stories make sense of personal experiences: Motives that shape autobiographical narratives. *Personality and Social Psychology Bulletin, 20*, 676–690.
Bing, M. N., Stewart, S. M., Davison, H. K., Green, P. D., McIntyre, M. D., & James, L. R. (2007). An integrative typology of personality assessment for aggression: Implications for predicting counterproductive workplace behavior. *Journal of Applied Psychology, 92*, 722–744.
Bluck, S. (2003). Autobiographical memory: Exploring its functions in everyday life. *Memory, 11*, 113–123.
Bluck, S., & Habermas, T. (2000). The life story schema. *Motivation and Emotion, 24*, 121–147.
Clark, K., & James, K. (1999). Justice and positive and negative creativity. *Creativity Research Journal, 12*, 311–320. doi:10.1207/s15326934crj1204_9.
Cropley, D. H., Kaufman, J. C., & Cropley, A. J. (2008). Malevolent creativity: A functional model of creativity in terrorism and crime. *Creativity Research Journal, 20*, 105–115. doi:10.1080/10400410802059424.

184   *The Ethics of Creativity*

De Dreu, C. W., & Nijstad, B. A. (2008). Mental set and creative thought in social conflict: Threat rigidity versus motivated focus. *Journal of Personality and Social Psychology, 95,* 648–661. doi:10.1037/0022-3514.95.3.648.

Ellspermann, S. J., Evans, G. W., & Basadur, M. (2007). The impact of training on the formulation of ill-structured problems. *Omega, 35,* 221–236.

Feist, G. J. (1998). A meta-analysis of personality in scientific and artistic creativity. *Personality and Social Psychology Review, 2,* 290–309.

Fontenot, N. A. (1993). Effects of training in creativity and creative problem finding upon business people. *Journal of Social Psychology, 133,* 11–22.

Furnham, A., & Nederstrom, M. (2010). Ability, demographic and personality predictors of creativity. *Personality and Individual Differences, 48,* 957–961.

Getzels, J. W., & Csikszentmihalyi, M. (1976). *The Creative Vision: A Longitudinal Study of Problem Finding in Art.* New York: John Wiley & Sons.

Goodstadt, B. E., & Hjelle, L. A. (1973). Power to the powerlessness: Locus of control and the use of power. *Journal of Personality and Social Psychology, 27,* 190–196.

Harris, D. J., & Reiter-Palmon, R. (2013). The influence of problem construction, implicit aggression, and task valence on malevolent creativity. Unpublished manuscript.

Harris, D. J., Reiter-Palmon, R., & Kaufman, J. C. (2013). The effect of emotional intelligence and task type on malevolent creativity. *Psychology of Aesthetics, Creativity, and the Arts, 7,* 237–244. doi:10.1037/a0032139.

Holyoak, K. J. (1984). Mental models in problem solving. In J. R. Anderson & S. M. Kosslyn (eds.), *Tutorials in Learning and Memory* (pp. 193–218). San Francisco, CA: Freeman.

Holyoak, K. J., & Thagard, P. (1997). The analogical mind. *American Psychologist, 52,* 35–44.

House, R. J., & Howell, J. M. (1992). Personality and charismatic leadership. *Leadership Quarterly, 3,* 81–108.

Howell, J. M., & Avolio, B. J. (1992). The ethics of charismatic leadership: Submission or liberation? *The Executive, 6,* 43–54.

Hunter, S. T., Bedell, K. E., & Mumford, M. D. (2007). Climate for creativity: A quantitative review. *Creativity Research Journal, 19,* 69–90.

Illies, J. J. & Reiter-Palmon, R. (2008). Responding destructively in leadership situations: The role of personal values and problem construction. *Journal of Business Ethics, 82,* 251–272.

James, K., Clark, K., & Cropanzano, R. (1999). Positive and negative creativity in groups, institutions, and organizations: A model and theoretical extension. *Creativity Research Journal, 12,* 211–226. doi:10.1207/s15326934crj1203_6.

James, L. R. (1998). Measurement of personality via conditional reasoning. *Organizational Research Methods, 1,* 131–163.

James, L. R., & LeBreton, J. M. (2011). Assessing the implicit personality through conditional reasoning. Washington, DC: American Psychological Association.

Johnson-Laird, P. N. (1983). *Mental Models: Towards a Cognitive Science of Language, Inference, and Consciousness.* Cambridge: Cambridge University Press.

Kay, D. S. (1991). Computer interaction: Debugging the problems. In R. J. Sternberg & P. A. Frensch (eds.), *Complex Problem Solving: Principles and Mechanisms* (pp. 317–340). Hillsdale, NJ: Lawrence Erlbaum Associates.

Keller, J. (2003). Parental images as a guide to leadership sense making: An attachment perspective on implicit leadership theories. *Leadership Quarterly, 14*, 141–160.

Krasikova, D. V., Green, S. G., & LeBreton, J. M. (2013). Destructive leadership: A theoretical review, integration, and future research agenda. *Journal of Management.* Advanced online publication. doi: 10.1177/0149206312471388.

Lee, S., & Dow, G. (2011). Malevolent creativity: Does personality influence malicious divergent thinking? *Creativity Research Journal, 23*, 73–82.

Ligon, G., Harris, D. J., & Hunter, S. T. (2012). Quantifying leader lives: What historiometric approaches can tell us. *Leadership Quarterly, 23*, 1104–1133. doi:10.1016/j.leaqua.2012.10.004.

Ligon, G. S., Hunter, S. T., & Mumford, M. D. (2008). Development of outstanding leadership: A life narrative approach. *Leadership Quarterly, 19*, 312–334.

McAdams, D. P. (1998). The role of defense in the life story. *Journal of Personality, 66*, 1125–1146.

McAdams, D. P. (2001). The psychology of life stories. *Review of General Psychology, 5*, 100–123.

McClelland, D. C. (1975). *Power: The Inner Experience.* New York: Irvington.

Mumford, M. D., Reiter-Palmon, R., & Redmond, M. R. (1994). Problem construction and cognition: Applying problem representations in ill-defined domains. In M. Runco (ed.), *Problem Finding, Problem Solving, and Creativity* (pp. 3–39). Norwood, NJ: Ablex.

Mumford, M. D., Connelly, M. S., Scott, G., Espejo, J., Sohl, L. M., Hunter, S. T., et al. (2005). Career experiences and scientific performance: A study of social, physical, life, and health sciences. *Creativity Research Journal, 17*, 105–129.

Mumford, M. D., Mobley, M. I., Uhlman, C. E., Reiter-Palmon, R., Doares, L. M. (1991). Process analytic models of creative thought. *Creativity Research Journal, 4*, 91–122.

Mumford, M. D., Supinski, E. P., Baughman, W. A., Costanza, D. P., & Threlfall, K. V. (1997). Process based measures of creative problem-solving skills; overall prediction. *Creativity Research Journal, 10*, 73–85.

O'Connor, J. A., Mumford, M. D., Clifton, T. C., Gessner, T. E., & Connelly, M. S. (1995). Charismatic leadership and destructiveness: A historiometric study. *Leadership Quarterly, 6*, 529–555.

Okuda, S. M., Runco, M. A., & Berger, D. E. (1991). Creativity and the finding and solving of real-world problems. *Journal of Psychoeducational Assessment, 9*, 45–53.

Pillemer, D. B. (2001). Momentous events and the life story. *Review of General Psychology, 5*, 123–134.

Reiter-Palmon, R., & Illies, J. J. (2004). Leadership and creativity: Understanding leadership from the creative problem-solving perspective. *Leadership Quarterly, 15*, 55–77.

Reiter-Palmon, R., Mumford, M. D., & Threlfall, K. V. (1998). Solving everyday problems creatively: The role of problem construction and personality type. *Creativity Research Journal, 11*, 187–197.

Reiter-Palmon, R., Mumford, M. D., Boes, O. J., & Runco, M. A. (1997). Problem construction and creativity: The role of ability, cue consistency, and active processing. *Creativity Research Journal, 10*, 9–24.

Rosenthal, S. A., & Pittinsky, T. L. (2006). Narcissistic leadership. *Leadership Quarterly, 17*, 617–633.

Rostan, S. M. (1994). Problem finding, problem solving and cognitive controls: An empirical investigation of critically acclaimed productivity. *Creativity Research Journal, 7*, 97–110.

Scott, G., Leritz, L. E., & Mumford, M. D. (2004). The effectiveness of creativity training: A quantitative review. *Creativity Research Journal, 16*, 327–361.

Sein, M. K., & Bostrom, R. P. (1989). Individual differences and conceptual models in training novice users. *Human–Computer Interaction, 4*, 197–229.

Simonton, D. K. (1983). Intergenerational transfer of individual differences in hereditary monarchs: Genetic, role-modeling, cohort, or sociocultural effects? *Journal of Personality and Social Psychology, 44*, 354–364. http://dx.doi.org/10.1037/0022-3514.44.2.354.

Simonton, D. K. (1994). *Greatness: Who Makes History and Why*. New York: Guilford Press.

Voss, J. F., Wolfe, C. R., Lawrence, J. A., & Engle, R. A. (1991). From representation to decision: An analysis of problem solving in international relations. In R. J. Sternberg & P. A. Frensch (eds.), *Complex Problem Solving: Principles and Mechanisms* (pp. 119–158). Hillsdale, NJ: Lawrence Erlbaum Associates.

Ward, T. B., Smith, S. M., & Finke, R. A. (1999). Creative cognition. In R. J. Sternberg (ed.), *Handbook of Creativity* (pp. 189–212). Cambridge: Cambridge University Press.

Wigert, B. G., & Reiter-Palmon, R. (2013). The influence of goals and constraints on creativity. Unpublished manuscript.

10
Intelligent Decision-Making Technology and Computational Ethics

Anthony Finn
University of South Australia, Australia

Hundredfold increases in performance per decade for microprocessors and similar levels of improvement in software have led to an explosion in the number of practical applications now envisioned for autonomous and intelligent decision-making technologies (IDT). Over the past 50 years, this has seen a sea change in miniaturization, maturation, diversity, and commercial availability of components needed to automate an array of systems successfully. Although expectations are yet to translate into the necessary critical mass to deliver against the many predictions made for the sector, IDT clearly offer considerable promise. In the same way as mobile phones, laptops, and the internet have recently transformed our daily lives, robotics and IDT are set to do likewise.

Our future thoughts of such systems increasingly rely on an extended idea of autonomy in which greater amounts of "original" decision-making are undertaken by the IDT. Although the technology is seen as overwhelmingly empowering society and working with—rather than in isolation from—humans, we are no longer confronted by the comforting paradigm that acts of innovation and creativity in a system's design, construction, decision execution, and physical action are undertaken by humans. We face a future in which sophisticated integration, optimization, and evolutionary software techniques combine with advanced fabrication procedures, hardware programming, and smart actuation to deliver IDT that synthesize code to execute on processors and machines *designed, programmed, and built by other machines*. These IDT will interact intelligently with their surroundings, abstracting information from sensed data, fusing this with information about their state to complete higher-level functions to plan and execute tasks intelligently, adaptively learning from environmental cues over time. The locus of innovation and decision between human and machine will become blurred.

This progressive excision of the human from the core design, decision-making, and physical tasks associated with systems means IDT will no longer be mere tools, which cannot act and think independently. They will be devices that, while they serve society in ways that overcome human frailties and limitations, will have an inherent "creativity" and hence "unpredictability" that relieve their users from certain responsibilities. In part, this can be justified on the basis that errors will no longer be a function of just humans, but of the inaccurate or ineffective facilitation of user understanding of how machines synthesize the technologies and/or the framework for, and execution of, their decision-making.

This does not imply that these IDT have sentience, intentionality, or will: they will be real-world devices controlled by algorithms. However, they will be called on progressively to perform as functional replacements for human alternatives, which will require sophisticated reasoning strategies and innovative behavior approaching that of people. In making this statement, it is important to note that the primary technological goal is unlikely to be delivery of the most interesting, realistic human imitation possible. Social and cultural drivers will undoubtedly shape the form many systems take, and anthropomorphic systems will continue to be studied and developed. However, commercial imperatives mean the form in which these advanced IDT emerge will be a compromise between what is technologically achievable, contextually appropriate, and economically desirable.

There are currently many limitations that make the aspiration of imaginative or "human-like" reasoning very unrealistic. Computer processing cannot currently mimic the processing capacity of the brain, the relevant architecture is not yet sufficiently well understood or optimized to allow complex perception or high-level reasoning, and the software cannot yet imitate the contextual decision-making capabilities of humans. There are, however, many examples where machines exceed human capabilities: we may parse and understand natural language better than IDT, but cannot hold a candle to an IDT that ranks and recommends information derived from billions of web pages. Similarly, the chess-playing computer Deep Blue famously played and beat the chess champion Gary Kasparov in 1997. This was followed by Deep Junior in 2003. For $50, you can now buy and run on a PC software programs that will easily defeat most grandmasters. What is perhaps most interesting is now that computers can regularly beat humans, we consider them chess players; and while such machines may once have received their chess-playing abilities from human programmers, this has long since been surpassed by a combination of raw computational power, huge memory, and good heuristics.

IDT are also trusted to make split-second decisions with life-or-death consequences because they perform repetitive, unpleasant, or dangerous tasks faster, more effectively, precisely, or efficiently than their human counterparts. Autopilots, driverless trains, combat systems, and "fire-and-forget" weapons are all examples of IDT that act independently of human control, albeit (in the last case) only to the extent that they manage mid-course and final trajectories, acquiring targets on the basis of certain limited stimuli. The ethical dimensions of final target designation, military value, and risk assessment are undertaken by humans, who ultimately have control of the weapon. As a result, a "safe" fully autonomous weapon has not yet been formally certified for operational use because the ability, for instance, to balance the key principles of the law of armed conflict without human intervention has not been demonstrated, and some say never should be.

If we were able to develop such a technology, at one level it could be considered equivalent to its fire-and-forget counterpart, with one being tasked by a silicon-based architecture on board the weapon and the other by a carbon-based one remote from it. However, only carbon-based architectures can be held accountable in law for their actions. Aside from the technological shortcomings, it is the intuitive implausibility of holding a machine to account that, for some, is sufficient to preclude them ever being allowed to make any final decision regarding lethal use of force. When it comes to creating such an IDT, there are two key issues: what is considered ethically acceptable and what is technically achievable.

In a study conducted by Moshkina and Arkin (2007), for which they canvassed opinion on the use of lethality by autonomous weapons, initial responses to the first of these questions are presented. The survey demographic spanned the general public, researchers in the field of robotics, policymakers, and military personnel. The results note that ethics are relative: an act carried out by a machine may be morally acceptable to a programmer, but perceived as immoral by indigenous cultures that witness its actions in theatre (Arkin, 2009).

In regard to the second question, the proposition of programming any machine, let alone a weapon, such that it is capable of making ethical decisions, is clearly audacious. It is attractive to dismiss as fantastical the notion that technology could ever perform as well as, or even outperform, a human. However, it is instructive to consider the findings of the Surgeon General's Office (2006), again noted by Arkin (2009). The findings were taken from a sample of 1,320 US soldiers and 447 US marines, and noted that only 47 percent of soldiers and 38 percent of marines agreed that non-combatants should be treated with dignity and respect. Similarly, the report noted that combat experience—particularly the loss of a colleague—was related to an increase in ethical violations. It is hard to believe that these figures cannot be improved on, although it remains an open question whether they can successfully carry out such tasks all of the time. Doing so challenges humans.

Such matters, of course, have held human fascination for over a century. Even though it risks sounding as though we are crossing into the realms of science fiction, as autonomous systems evolve and deal with increasingly complex and dynamic environments, pressure to develop machines with the ability to resolve challenges with an ethical dimension will increase. Furthermore, several research groups are devising technological solutions to contextual interpretation of circumstances from the perspective of what IDT *ought* to do rather than simply allowing it to do *what it may* (see Anderson, Anderson, & Armen, 2006; Arkin, 2009; Anderson & Anderson, 2011; Lin, Abney, & Bekey, 2012; Moor, 2006; Wallach & Allen, 2009).

There are several ways in which we currently ensure that machines behave ethically. For instance, a machine's actions and decisions may be constrained so that unethical outcomes are avoided. Or software can be created explicitly to contain ethical maxims so that the machine implicitly promotes ethical behavior. Many machines already do this by ensuring adherence to legal principles and regulations: autopilots address critical safety, regulatory, and reliability concerns; and automated bank tellers and internet banking agents attempt to transfer only correct amounts of money. These machines are carefully constructed and do not simply have algorithms that tell them to "behave honestly."

More complex machines that might be considered to "act ethically" include those involved in disaster relief (Anderson & Anderson, 2011). In a major emergency, such as a hurricane, earthquake, or tsunami, networks of sensors and computers are used to track and process information about who needs the most help and how such relief is best distributed. In the extreme, such decisions effectively determine who lives and who dies. Some might say that only humans should be permitted to exercise such judgments. But if IDT outperform human decision-making (for example, in terms of lives saved), we have a solid ethical basis for letting the IDT take these decisions. Similar situations exist for soldiers on the front line who might benefit from weapons with automated target discrimination and response options; these technologies may save own-force lives.

The IDT discussed in the rest of this chapter, however, have the highest-order software agents: those capable of making and reasonably justifying explicit ethical judgments. Such machines conceptually might be used to control autonomous weapons; as ethics tutors or coaches for young doctors, soldiers, and lawyers; as decision aids in a person's right to die (for example, when faced with crippling injury or illness); as specialist decision aids for biomedical researchers or triage specialists following large-scale emergencies; or even a social aid suggesting when to tell the truth (McLaren, 2006).

This is not to suggest that the IDT will *know* it is making ethical judgments. It will treat ethical decisions the same as any other decisions. In fact, instantiation of this reasoning, while complex, will "simply" consist of some combination of law application, constraint application, reasoning by analogy, planning, and optimization (McDermott, 2011). Ethical reasoning by IDT can be considered prioritizing and deciding whether a context is sufficiently similar to circumstances envisaged for the ethical premise to be applicable (Moor, 2011). That is, the agent needs only to balance ethical propositions and behave ethically to deliver "artificial ethics."

Challenges for the development of ethical IDT

The challenge of delivering a machine capable of making ethical judgments "will involve a broad range of engineering, ethical and legal considerations and a full understanding will require a dialogue between legal theorists, engineers, computer scientists, philosophers, developmental psychologists and other social scientists" (Moor, 2006). Still, in regard to developing "everyday" computational ethics, a number of authors (Anderson & Anderson, 2011; Arkin, 2009; Borenstein, 2008; Brngsjord, Arkoudas, & Bello, 2006; Freedy et al., 2007; Hendler, 2006; McLaren, 2011) have described the importance, approaches, issues, visions, and state of the art for the technology. Generally, these authors acknowledge that we neither want nor need the IDT to derive its own code of beliefs from first principles. Rather, we wish it to apply principles derived previously by humanity and coded within preexisting frameworks.

While clearly daunting—and many may consider such thoughts of creating artificial morality rather far-fetched—a number of practical

192 *The Ethics of Creativity*

frameworks for governing the ethical control response of IDT have been proposed (Allen, Wallach, & Smit, 2006; Anderson, Anderson, & Armen, 2004, 2005, 2006; Anderson & Anderson, 2011; Arkin, 1998, 2009; Borenstein, 2008; Bringsjord, Arkoudas, & Bello, 2006; Canning, 2006; Chopra & White, 2004; Karnow, 1996; Lindsey & Pecheur, 2003; Moor, 2006; Powers, 2006; Wiegel, 2006). Even so, this will tax us in a number of ways, which include:

- The creation of a "mechanical conscience" will challenge our understanding of morality to its extreme, as "we must formally express our knowledge of any moral framework in a manner that engineers and computer scientists can understand and express in software and hardware" (Wiegel, 2006).
- Many ethical theories are not well understood even by specialists in the field, let alone software designers and engineers. Consequently, determination of a value proposition is complex (Studer et al., 2006).
- Ethical reasoning is based on abstract principles, which often conflict with each other in specific situations. If more than one principle applies, it is often not clear how to resolve the matter, as first-order logic is not usually available (McLaren, 2006).
- Many laws and ethical frameworks are intended to be self-consistent and complete. They provide a broad framework, on which we can agree, as a basis for the development of agents. However, the premises, beliefs, and principles that humans use to make the decisions linked to ethics are varied, and the interpretations of many ethical tenets are not yet agreed on (Anderson & Anderson, 2011). The conditions, premises, or clauses that would frame the development of software are hence imprecise, are subject to interpretation, and may have different meanings in different contexts (Arkin, 2009).
- If an ethical agent is to prevent an IDT from acting in some unethical fashion, it also may need the capacity to explain to its human supervisor the underlying reasons for its logic in case the human wishes to override it (Anderson, Anderson, & Armen, 2004).
- The tasks involving the IDT may need to rely on reasoning and judgment that are linked to the execution of higher-order obligations. For instance, autonomous weapons may need to fit within a military command-and-control hierarchy. These higher-order tasks are more difficult to deal with and understand as the prioritization of goals and ordering of subtasks must be derived directly or by implication from overarching strategies (Brngsjord, Arkoudas, & Bello, 2006).

- Problems that are complex for humans are less well understood, less structured, and harder to decompose into definable components. There is a high likelihood of ambiguity and the likelihood that one decision will have an impact on a subsequent one.
- From time to time, it is necessary to reframe a difficult problem rather than interpreting it within the existing problem frame. Recognizing this and dealing effectively with complex novelty (and discriminating it from difficulty) require a strong sense of self-awareness.
- The measurement process by which we observe the key criteria, then distill the large set of metrics available to an efficient model or group for any given evaluation, has yet to be refined.

One further caution is that, while IDT may be trained against data selected by their owners and operators, and we would expect the IDT to be trained in a way that leads to improvements, there may be less scrupulous individuals who seek to identify and exploit the IDT equivalent of a fanatic, a system that overrides some preordained taboo on the basis of some learned behavior (Arkin, 2009).

Legal challenges for ethical IDT

In early phases, it seems likely that legislators will insist IDT are used in a particular way, operated under particular conditions, or have certified users. Eventually, we will need to address what failure rate and what type of failures we are willing to accept and how safety requirements should be traded off against criteria such as cost and performance. Typically, these cost–capability decisions will be made on the basis of some statistically significant criteria such as "the life expectancy of a human shall not be altered by using such a system." Unfortunately, such criteria do not provide us with any absolute measure of what constitutes *safe* or *known malfunction*.

Clearly, the system will need to be designed to "world's best practice." But this on its own is probably insufficient, not least because such standards and practices are currently informal and therefore not legally binding. If an IDT is involved in an accident, the issue of how it was used is of interest. However, it may be extremely difficult to say whether this was inappropriate; often, the best that can be concluded is that the technology or its human–machine interface was used in a particular way and specific consequences followed. This may be particularly true for complex environments, where the only data gathered and usable in regard to identifying the cause may come from the IDT itself, which, of course, may be designed not to allow information to be disclosed that may "incriminate" it (the allocation of responsibility is an important legal principle, so jurisdictions try to ascribe liability by tracing the vector of causation back to the agency where mistakes were made. If, as is suggested later, IDT are provided with separate legal identity, then intentionally not providing "self-incriminating" evidence is considered legally sound practice in many jurisdictions).

To understand these challenges in depth, we need to examine texts such as Solum (1992), Sparrow (2007), Asaro (2007b), and Finn & Scheding (2010). Broadly, however, the candidates on whom responsibility for accidents or infringements will fall include putting it down to an unfortunate mistake, or ascribing it to the user, acquirer, owner, manufacturer, integrator, designer, programmer, designer, engineer, and so on. There are arguments both for and against each, but these may largely be distilled to a few key issues. For example, inherent in the nature of such technology is its ability to make choices other than those directly programmed into it. Similarly, the decision-making capability of the IDT implies that instructions issued by users "influence but do not prescribe" its actions (Sparrow, 2007); and the internal states of the IDT are usually so complex that the actions of the machine are indeterminable. Moreover, the technology will likely learn from its environment or mistakes, which will make reproducible and verifiable testing very difficult if not impossible.

Each candidate will also know that the IDT could operate erroneously or that the IDT could develop unresolved or contradictory states; the system will not be released untested. As lawyers are employed to provide advice on responsibilities in producing, advertising, selling, and using products—and companies that make ethical agents are unlikely to be any different in this regard—it seems likely that the candidates will adequately fulfill their legal obligations in regard to any potential issues of product liability. Thereafter, assuming that there was not a faulty implementation of an otherwise acceptable design, blaming any of the candidates would seem to be both unfair and impractical.

We also face the challenge of accounting for the polymorphic nature of the IDT—which will interpret data from a number of different perspectives and manipulate information in accordance with environmental conditions, the nature of the stated objectives, and the problem at hand—a challenge that will be compounded as the decision-making elements will likely be distributed across a number of modules, processors, and concurrently interacting components (machine and humans), none able to make the decision in its own right. There will be interaction between multiple programs developed by multiple programmers (possibly from different software houses) on a variety of processors, operating systems, and architectures (perhaps unknown to each other in advance) and possibly distributed across a network of processors. The end result is that determining any human vector of causation will prove extremely challenging.

We must also acknowledge that the failure of such systems may not always be due to human negligence during their creation, acquisition, or operation, particularly if modern design paradigms are employed. That is, the layout of a processor board was automated; a machine learning algorithm adaptively modified key coefficients; or code testing, development, or debugging were the result of evolutionary software techniques (Arcuri & Yao, 2008). As a result, it may be beneficial to avoid trying to identify a causal relationship between accidents and the functional component level of the IDT. That is, given the difficulties in attributing responsibility to the humans or their agencies, why not ascribe it to the IDT?

To address this issue, we must first consider the legal concept of agency, which is a highly specialized field of law, but it may be summarized as follows: an agent is empowered by its principal to negotiate and make various arrangements on its behalf; thereafter, the principal is bound by the contract that its agent signs as if the principal itself had signed, unless it is possible to prove misconduct. Moreover, the agent's actual authority extends to cases of apparent authority, where the agent has no actual authority but where the principal permits him to believe that he has authority (Asaro, 2007b). To this end, as IDT become more sophisticated and are able to take a range of ethical decisions on behalf of their human counterparts, it is attractive to think of the technology as agents of their user. It is recognized that the law may hesitate to assign responsibility to an IDT when it could more reasonably have been ascribed to someone who might have been able to control the outcome. However, it may be preferable to the alternatives, particularly if the best that can be achieved is costly, complex, indeterminate, or unpredictable.

The usual philosophical and legal debate over whether or not legal personality[1] can be ascribed to artificial agents centers around the list of necessary and sufficient conditions that must be met by the artificial agent in order for it to be recognized as an equivalent to a human (for example, Allen, Wallach, & Smit, 2006; Asaro, 2007a; Chopra & White, 2004; Kerr, 2004; Solum, 1992; Teubner, 2007). While machines may ultimately achieve this status, it is not likely to occur any time soon. Consequently, we may choose to afford IDT quasi-legal status and allow them responsibility for a limited set of decisions and actions.[2] In this way, we might consider sophisticated IDT to hold diminished responsibilities such that the liability for certain decisions could be transferred to them or differentially apportioned between IDT and humans.

Clearly, in order to consider IDT as agents, we must imagine some form of "contractual" obligation to exist between the user and the IDT. The mechanism through which electronic agents are able to create contracts is typically known as "attribution law" (Kerr, 2004), which acknowledges the ability of electronic agents to conclude contracts independently of human review and alter the rights and obligations of their principals in their absence. In essence, this law stipulates that a person's actions include those taken by human agents and those taken by electronic ones. In other words, any transaction entered into by an electronic agent is attributed to the person using it, the requisite intention flowing from the programming and use of IDT (Teubner, 2007).

There appear to be five options that may be used as a legal framework: three that require us to apply or slightly modify our interpretation of "current legal doctrine," and two that adopt more radical approaches. The first three options entail treating IDT as tools, insisting on human intervention, and achieving technical equivalence. The last two, treating IDT as agents and separate legal personality, are briefly discussed here. All are discussed in greater detail in Finn and Scheding (2010).

IDT as agents

Treating IDT as agents requires us to treat the matter as if the use of an ethical agent were akin to a "contract" between two parties, with the IDT acting as an agent on the supervisor's behalf. In the same way that a person can be bound by signing an unread contract—or even a contract generated by a computer over which the principal has no direct control—we assume that the human may be bound by the decisions of the IDT acting in accordance with his general intentions even though he was not cognizant of the detail of its actions in terms of any specific commitment. This option, however, requires us to distinguish between autonomy and unpredictability.

In the law of agency, a party's assent is not necessary to form a contract. It is sufficient, judged according to "standards of reasonableness," for one party to believe that the other party intended to agree; the real but unexpressed state of the other party's mind is irrelevant. In other words, at the point of use, a human will have certain expectations of the way in which the IDT will behave. In our case, this would mean that we accept that, without ever "knowing the mind" of the IDT, a "contract" between the user and the IDT is assumed. If the IDT behaves autonomously, but not erratically or unpredictably, responsibility for its actions would then be in accordance with the current interpretation of the law. Alternatively, if the behavior of the IDT is deemed erratic or unpredictable, then it would be acknowledged that liability for its actions does not fall on the users.

While attractive, the practical limitations of this approach are significant. For instance, the distinction between expected autonomous and erratic behavior must be determined. That is, standards of reasonable behavior must be established and defined for all IDT capable of making such decisions. To achieve this goal, it is not enough simply to determine statistical norms. We will need to compare these to previously agreed metrics that represent (say) human behavior; and most would agree that these either do not currently exist or have not yet been universally agreed. For this, we must establish which criteria define human equivalence, how many criteria are essential as opposed to merely useful, and what we do if the thresholds for several criteria are exceeded, but not others.

Separate legal personality

IDT do not appear to fall within any of the existing legal principles, so why not create an exception? To do this, we need to accept that human intention need not underlie the use of an IDT, making an ethical deliberation on our behalf. In other words, we need to assume that the user's generalized and indirect intention is sufficient to render the use of IDT legal. While reliant on the technical competence of the IDT to achieve a certain practical standard (for example, human equivalence), this would, in effect, need the IDT to hold a separate legal personality.

There are obviously a number of objections to non-humans holding legal status and rights, and some simply will dismiss the notion of agency or a separate legal personality for an inanimate object as fantastical and not something deserving serious consideration. Holding separate legal status does not imply that the IDT is considered a person; although the technical legal meaning of *person* is a "subject of legal rights and duties" (Al Majeed, 2007). It simply recognizes that being human is not a necessary condition of being accorded legal personality. Furthermore, there are well-established precedents for dealing with abstract and inanimate objects that enjoy legal rights and duties (for example, nation-states, churches, gods, shrines, corporations, ships, dead people, trees, and animals). There are also theories that define the legal criteria for agency and identity, not on the basis of ontological properties (mind, soul, empathy, and the reflexive capacities) possessed by an entity, but on the basis of the entity's ability to communicate (Teubner, 2007). Indeed, a ship may be the subject of a proceeding and can be found "guilty"; and between the ninth and nineteenth centuries, there were over 200 documented cases of animals and "miscellaneous vermin" being put on trial for a range of crimes and misdemeanors. Perhaps the most famous case was the rats of Autun who were put on trial for destroying crops. They won their case, albeit with skills of a local lawyer, Barthelemy Chassenee (Ewald, 1995).

Many commentators have discussed artificial intelligence and software agents holding rights (Asaro, 2007b; Chopra & White, 2004; Finn & Scheding, 2010; Kerr, 2004; Solum, 1992; Teubner, 2007). The major philosophical objections may be summarized as follows: it is only right that natural persons be given legal rights; IDT lack soul, consciousness, intentionality, free will, and so on; as human creations, IDT can never be more than human property; such approaches are not necessary as other solutions are adequate; the inability to sue an errant IDT is not, in practice, a significant loss; IDT have no assets, so any judgment against them is meaningless; and, in multiagent systems that are in communication with one another and comprise multiple copies of the same code, it is not clear whether the entity should be recognized as singular or plural in a legal sense.

Here we simply note that there are clearly contrary intuitions and arguments, and each depends on values or assumptions that are not necessarily universally shared (for example, is a child or a human clone a human creation?). Furthermore, if consciousness is a computational property of the neuronal structure and carbon-based architecture of the brain, and we can model the brain (Markram, 2006), why do we consider consciousness to be unreproducible in the silicon-based architectures of a suitable computer?[3] Similarly, if an IDT capable of adaptive learning is examined at the level of its interaction with the environment, it will appear autonomous, interactive, and adaptive. On the other hand, if the code is examined, it will "disappoint," being shown simply to be following algorithmic rules. So it is for the levels of abstraction in the brain, except those levels are of the neuronal and the conscious.

Perhaps the most potent objection to an IDT "taking responsibility" is the question: "Is it the real decision-maker?" A prerequisite for such status is an ability to make such decisions, and we have established by technological extrapolation that this (eventually) will be the case, noting the complexity of identifying the locus of decision-making. The key argument centers on whether the natural decision-maker (the human), making the decision to use the IDT in the first place, constitutes *the* crucial discretionary judgment, as the power to make such decisions essentially identifies the principal authority.

Assuming that IDT are employed because they are more effective, efficient, or consistent than their human counterparts, how can we reasonably argue that the real decision-maker is its user? The user may only be making a few discretionary interventions on behalf of the IDT, leaving the technology to cope with the majority of decisions, but with the overall decision-making consensus improved. Identifying the human's decision to use IDT as the central issue then poses a series of questions: When should the user elect to rely on the IDT's judgment and when their own? Should these decisions to intervene be based on a formal framework of metrics, the relative performances of users and IDT, or less rational factors? What should the consequences be of inappropriately intervening (or not), if the same choice had previously resulted in a beneficial outcome? What might any mitigating circumstances be?

Conclusion

There are many level-headed people who will consider it futile to discuss abstract concepts when we have trouble making relatively simple machines work in the real world. However, such discussions are valuable as they provide a way of thinking about IDT development from a different perspective and help to avoid it outpacing our current legal, ethical, and regulatory regimes. Implicit in this suggestion, of course, is that, if only lawyers, politicians, and ethicists thought more about these topics, our law and ethical frameworks could keep up. To some extent this is true: we should better anticipate and accommodate technological change as this allows us to develop an understanding of existing and potential options in the light of certain hypothetical possibilities. These analyses also permit the creation of IDT with appropriate characteristics designed into the way they operate. Unfortunately, novelty in the technology, difficulty in accurately forecasting consequences and implications, and an unwillingness to burden agencies with additional work mean our frameworks typically lag behind.

At present, IDT are ethically oblivious to the dilemmas they face: the weapon does not understand its target is a barracks, a school, or a hospital; the out-of-control driverless car does not know that swerving to the left will kill a child, and to the right, a mother. It is also reasonable to ask: "If we were able to develop such technology, would we want to?" It is, of course, easy to dodge the question by asserting that these goals are lofty and unachievable. However, investment of hundreds of billions of dollars each year into hardware improvements, and better formulation of the mathematics of problems, will continue to bring gains. These will allow us to solve given problems more quickly and more complex problems in the same amount of time. The benefits of such advances almost certainly will outweigh the costs. However, this optimism about the future is, like so many things, not a free lunch. We cannot simply sit back and hope for the best. Inexorably, as IDT become more human-centric and their decision-making more creative, we will be confronted by the challenges of how to code software capable of resolving complex dilemmas that have ethical dimensions.

Notes

1. Typically, a legal persona has the capacity to sue and be sued and to hold property in its own name.
2. Many legal entities (for example, children, corporations, and the mentally impaired) frequently act through agents, and not all legal entities share the same rights and obligations: some (for example, marriage) depend on age, whereas others (for example, voting and imprisonment) are restricted to humans.
3. One possible answer is that, according to Penrose (1989), the physiological processes underlying thought may involve the superposition of a number of quantum states, each of which performs a key calculation, before the differences and distributions in mass and energy cause the states to collapse to a single, measurable state. This process cannot be replicated by any computer yet conceived.

References

Al-Majeed, W. (2007). Electronic agents and legal personality. BILETA Annual Conference, Hertfordshire, 16–17 April.

Allen, C., Wallach, W., & Smit, I. (2006). Why machine ethics? *IEEE Journal of Intelligent Systems, 21*(4), 12–17.

Anderson, M., & Anderson, S. (2011). *Machine Ethics*. Cambridge: Cambridge University Press.

Anderson, M., Anderson, S. L., & Armen, C. (2004). Towards machine ethics. *AAAI-04 Workshop on Agent Organisations: Theory and Practic*e, San Jose, CA, July 25.

Anderson, M., Anderson, S., & Armen, C. (2005). Towards machine ethics: Implementing two action-based ethical theories. *2005 AAAI Technical Report FS-05-06*, 1–7.

Anderson, M., Anderson, S. L., & Armen, C. (2006). An approach to computer ethics. *IEEE Journal of Intelligent Systems, 21*(4), 56–63.

Arcuri, A., & Yao, X. (2008) A novel co-evolutionary approach to automatic software bug fixing. *Proceedings of the 2008 IEEE Congress on Evolutionary Computation (CEC 2008)* (pp. 162–8). Hong Kong, China: IEEE Press.

Arkin, R. (1998). *Behavior Based Robotics.* Cambridge, MA: MIT Press.

Arkin, R. (2009). *Governing Lethal Behaviour in Autonomous Robots.* Boca Raton, FL: Chapman & Hall/CRC Press.

Asaro, P. (2007a). Robots and responsibility from a legal perspective. *Proceedings IEEE Conference on Robotics and Automation, Workshop on Robot Ethics,* Rome, April.

Asaro, P. (2007b). How just could a robot war be? Umeå University, Sweden, 2007.

Borenstein, J. (2008). The ethics of autonomous military robots. *Studies in Ethics, Law and Technology, 2*(1).

Bringsjord, S., Arkoudas, K., Bello, P. (2006). Ethical robots: The future can heed us. *IEEE Journal of Intelligent Systems,* 12–17.

Canning, J. (2006). A concept of operations for autonomous weapons. *NDIA Conference on Disruptive Technologies,* Washington, DC, 6–7 September.

Chopra, S., & White, L. (2004). Artificial agents: Personhood in law and philosophy. *European Conference on Artificial Intelligence,* IOS Press, 635–639.

Ewald, W. (1995). Comparative jurisprudence: What was it like to try a rat? *American Journal of Comparative Law, 143*(1), 889–990.

Finn, R. A., & Scheding, S. (2010). *Developments and Challenges for Autonomous Unmanned Vehicles.* Berlin: Springer.

Freedy, A., DeVisser, E., Weltman, G., & Coeyman, N. (2007). Measurement of trust in human–robot collaboration. *Proceedings of International Conference on Collaborative Technologies and Systems IEEE.*

Hendler, J. (2006). Computers play chess, humans play go. *IEEE Journal of Intelligent Systems, 21,* 2–3.

Karnow, C. (1996). Liability for distributed artificial intelligence. *Berkeley Technical Law Journal, 11,* 149–174.

Kerr, I. (2004). Bots, babes and the Californication of commerce. *University of Ottawa Law and Technology Journal, 1,* 285–324.

Lin, P., Abney, K., & Bekey, G. (2012). *Robot Ethics: The Ethical and Social Implications of Robotics.* Cambridge, MA: MIT Press.

Lindsey, A., & Pecheur, C. (2003). Simulation-based verification of livingstone applications. *Proceedings of Workshop on Model-Checking for Dependable Software-Intensive Systems,* San Francisco, CA.

Markram, H. (2006). The blue brain project. *Nature Reviews of Neuroscience, 7,* 153–160.

McDermott, A. (2011). *Why Machine Ethics.* New York: Cambridge University Press.

McLaren, B. (2006). Computational models of ethical reasoning. *IEEE Journal of Intelligent Systems, 21*(4), 2–3.

McLaren, B. (2011). *Computational Models of Ethical Reasoning: Challenges, Initial Steps, Future Directions, in Machine Ethics.* New York: Cambridge University Press.

Moor, J. (2006). The nature, importance, and difficulty of machine ethics. *IEEE Journal of Intelligent Systems, 21*(4), 18–21.

Moor, J. (2011). The nature, importance, and difficulty of machine ethics. In M. Anderson & S. Anderson (eds.), *Machine Ethics* (pp. 13–20). New York: Cambridge University Press.

Moshkina, L., & Arkin, R. (2007). *Lethality & Autonomous Systems: Survey and Design Results*, Georgia Institute of Technology, Technical Report GIT-GVU-07-16.

Penrose, R. (1989). *Emperor's New Mind: Concerning Minds and the Laws of Physics.* Oxford: Oxford University Press.

Powers, T. (2006). Prospects for a Kantian machine. *IEEE Journal of Intelligent Systems, 21*(4), 46–51.

Solum, L. B. (1992). Legal personhood for artificial intelligence. *North Carolina Law Review, 70,* 1231–1287.

Sparrow, R. (2007). Killer robots. *Journal of Applied Philosophy, 24*(1), 62–77.

Studer, R., Ankolekar, A., Hitzler, P., & Sure, Y. (2006). A semantic future for AI. *IEEE Journal of Intelligent Systems, 21*(4), 8–9.

Surgeon General's Office. (2006). Mental Health Advisory Team (MHAT) IV Operation Iraqi Freedom 02–07, Final report.

Teubner, G. (2007). Rights of nonhumans, Max Weber Lecture 2007/04, European University Institute.

Wallach, W., & Allen, C. (2009). *Moral Machines: Teaching Robots Right from Wrong.* Oxford: Oxford University Press.

Wiegel, V. (2006). Building blocks for artificial moral agents. *Proceedings of the Tenth International Conference on Simulation and Synthesis of Living Systems,* Bloomington, IN.

Part III

What Role Does Ethics Play in Supporting or Thwarting Creativity?

The following five chapters examine how aspects of ethical constraints or decision-making affect creative work in five domains: arts, sciences, photography, education, and leadership.

11
Creative Transformations of Ethical Challenges

Vera John-Steiner and Reuben Hersh
University of New Mexico, USA

There is a widely held assumption that the appeal of novelty, the rewards of discovery, and the beauty of artistic expression provide their own rewards and constitute a claim for the human capacity for goodness. In celebrating these contributions to our lives, we seldom examine the possibility that such a creative contribution can also pose a serious threat to our very survival. In this chapter, we explore the ethics of choices that accompanied creative contributions in three different historical settings.

The creativity researcher Howard Gruber recognized the tension between "task-oriented" and what he called "world-oriented" motivational dynamics (Wallace & Gruber, 1989, pp. 284–285). Engagement in creative activities can be isolating from the social world that surrounds creative people because of the intensity with which such individuals pursue their passion. But that very intensity that is so central to writing, painting, or scientific work can also make these individuals sensitive to the great challenges we face as social beings. Gruber wrote of the claims the world makes on us, whether as looming environmental disasters, incessant wars, inequality, or pervasive fatal illness. As these challenges surround us, meeting them requires individual and social resources that go beyond habitual behaviors. They require empathy, vision, courage, and a view of the self interdependent with others.

In examining the motivational sources of creative work, Gruber also mentioned ego orientation or extrinsic motivation. It "refers to an attitude toward work that is motivated by desire for rewards not inherent in the task itself, rewards such as recognition, prestige, prizes, money, privileges and power. Task orientation, or intrinsic motivation, refers to an attitude toward work that is motivated by the intrinsic nature or demand-character of the task itself" (Wallace & Gruber, 1989, p. 284). Creative people who are driven by the satisfaction achieved through rewards for creative work are seldom oriented toward using their talents to improve their world. They can form temporary alliances with more visionary individuals, who experience the pain of others in a very personal and disturbing way. The latter are gratified when, through empathy, they can reach beyond the confines of their individual concerns. There is an interesting tension in people who find in their work an intense and pleasurable absorption, what Csikszentmihalyi (1990) has called "flow," but who can also devote themselves to causes beyond those that provide them with creative gratification. Gruber called these individuals "cases of 'creative altruism', in which a person displays extraordinary moral responsibility, devoting a significant portion of time and energy to some project transcending an immediate need and experience" (Wallace & Gruber, 1989, p. 285).

We present three case studies of individuals who demonstrate exemplary creative altruism, and examine the ways in which they respond to challenging circumstances. One example of creative altruism was the physicist Leo Szilard, who confronted one of the most difficult ethical choices faced by any individual in the twentieth century. Another such individual was the painter Friedl Dicker-Brandeis, who taught art to children in the 1940s in the Terezín concentration camp. Our third example is Clarence Stephens, an African American mathematics professor whose approach to teaching difficult subject matter was creatively to transform a traditional rigid approach to a humanistic one.

Leo Szilard

Morality and creativity have usually been considered independent domains, with little attention being paid to the moral implications of a creative innovation. But history has taught us otherwise: we are all still living with the consequences of moral choices made by great, creatively innovative scientists in the early and mid-twentieth century. In the following case study, we present the moral and ethical challenges faced by Leo Szilard, a man determined to confront the tragic consequences of his scientific discovery.

During the 1920s and 1930s, physics was experiencing a golden age, with major discoveries being made in subatomic studies. This work was conducted in universities by large communities of scientists drawn from many countries, including Berlin's Humboldt Institute, where Albert Einstein taught for a while. Many young scientists traveled between these centers and British laboratories, learning of new developments through face-to-face interactions. One of these was the Hungarian physicist Leo Szilard, who after years of work in different continental centers settled in England in 1933. It was Szilard who first conceived of the possibility of a chain reaction of nuclear fission—either to produce useful energy, or to commit extinction. He twice played a singular and unique role with respect to the greatest threat to the survival of the human race: the atomic bomb.

In nuclear fission, a heavy atom breaks in two when struck by a neutral elementary particle, a neutron. ("Neutral" means having no electrical charge, neither positive nor negative.) Some of the original mass would be converted into radiant energy, and more neutrons would be released. Szilard was the first person to realize that, in this way, a nuclear chain reaction could take place, releasing huge amounts of energy. The great release of energy is a consequence of Einstein's famous equation: $e = mc^2$.

Szilard first started to think about the chain reaction problem after reading a summary of a talk by Lord Ernest Rutherford, an eminent British physicist. Rutherford minimized the potential importance of nuclear fission as a source of large quantities of energy. Szilard was challenged by this prediction and spent large amounts of time trying to come up with a different view. He synthesized diverse research findings, such as that of the chain reaction studied by chemists including his Hungarian friend Michael Polanyi and Rutherford's own findings of "critical mass" (Lanquette, 1992, p. 134). Szilard's process of scientific creativity corresponded with many other documented cases (John-Steiner, 1985) where synthesis, analogy, and intellectual dialogue are basic features of discovery.

Szilard's insight that a nuclear bomb could be built based on the chain reaction occurred when the Second World War was looming. What in 1933 must have seemed very far out and unlikely to be realized suddenly became a terrifying threat in 1939. In their laboratory, two German physicists, Otto Hahn and Fritz Strassman, succeeded in splitting the nucleus of a uranium isotope, number 235. These physicists and their laboratory belonged to Adolf Hitler's Third Reich, the Nazi regime that threatened all of Europe. The ethical dilemma is clear. It is hard to imagine something worse than the whole world ruled by Adolf Hitler. If the atom bomb was going to come about, it was vital that it should be built by the USA and its allies before the Nazis could do so. The negative side was that the world would become much less safe for the human race once atom bombs were manufactured. Nevertheless, during wartime, with the dreadful fear of a Nazi victory predominant, this negative side was easy to minimize.

This was the first of two ethical dilemmas about the atom bomb that Leo Szilard would face. He quickly decided that he must motivate the USA to start working to build its own atom bomb ahead of the Germans. He was a foreign refugee, without any power or influence. Nevertheless, he succeeded with help from another refugee physicist who was so famous that a letter from him could not be ignored. Albert Einstein admitted to Szilard that the idea of a nuclear chain reaction had never occurred to him. Szilard composed a letter to President Franklin D. Roosevelt, a letter that Einstein signed, even though an ardent pacifist. Urging Roosevelt to use his own great discovery to create a weapon of mass destruction must have been hard indeed for Einstein to advocate. He later declared that he wished he had become a plumber rather than a physicist.

Szilard played a critical role in informing President Roosevelt of the feasibility of building a bomb, which became realized in the Manhattan Project. He spent the war working in the code-named "Metallurgical Laboratory" in Chicago, under Enrico Fermi, and there the first sustained chain reaction of nuclear fission was achieved. Then the action moved to Los Alamos, Hanford, and Oak Ridge. General Leslie Groves, the director of the Manhattan Project, considered Szilard not only a nuisance, but a security risk. He forbade Szilard from visiting Los Alamos, and even had him shadowed by FBI agents all through the war. Szilard tried at times to make friends with his FBI shadows. Their reports of his comings and goings make absurd reading today. A few years later, the war in Europe came to an end; there was no longer any fear of a Nazi bomb. The original motivation for the whole project had vanished. However, Hitler's Japanese allies continued to fight. Vast Russian armies were moved to the Far East in preparation for an invasion of Japanese-held Mongolia. Japan was trying to end the war without yielding an unconditional surrender. It was expected that invading Japan would cost the lives of many American soldiers. Should the bomb be dropped on Japan? Many in the scientific community, including Einstein and Szilard, opposed using "the bomb" on a civilian population, rather than demonstrating its power by dropping it on an uninhabited island.

Great efforts had been made to keep the Manhattan Project secret in order to maintain a monopoly of atomic weapons for the USA. Nevertheless, Szilard and other physicists knew that once an atomic explosion took place, telling the whole world that an atomic bomb was feasible, in only a few years the Soviet Union would also have atom bombs. What would then follow would be a nuclear arms race—the utmost danger, an existential threat to human survival. Szilard was well aware that this danger was, in part, his own creation and he would spend decades striving against it.

Using another letter from Einstein, Szilard scheduled a meeting with Eleanor Roosevelt for May 8, 1945. He planned to give her information that would caution President Roosevelt about the danger of a nuclear arms race, if the A-bomb was used before an international control agreement could be discussed with the Soviets. But on April 12, President Roosevelt died. Szilard's attempt to meet with President Truman led instead to a May 28, 1945 meeting with James Byrnes, who would soon become Secretary of State. However, Byrnes thoroughly disagreed with Szilard's views.

Other physicists shared Szilard's worries, including James Franck, Director of the Chemistry Division of the Metallurgical Lab at the University of Chicago. Szilard was one of the main authors of the Franck Report issued in June 1945, two months before the bombing of Hiroshima. This warned that even if the A-bomb helped save lives in this war, using it could lead to a nuclear arms race and a possible nuclear war. Here are some of the key passages:

> If no efficient international agreement is achieved, the race of nuclear armaments will be on in earnest not later than the morning after our first demonstration of the existence of nuclear weapons. (II. Prospectives of Armaments Race, Para 9)

> only lack of mutual trust, and not lack of desire for agreement, can stand in the path of an efficient agreement for the prevention of nuclear warfare. (III. Prospectives of Agreement, Para 1)

> It will be very difficult to persuade the world that a nation which was capable of secretly preparing and suddenly releasing a weapon, as indiscriminate as the rocket bomb and a thousand times more destructive, is to be trusted in its proclaimed desire of having such weapons abolished by international agreement. (III. Prospectives of Agreement, Para 3)

> From this point of view a demonstration of the new weapon may best be made before the eyes of representatives of all United Nations, on the desert or a barren island. (III. Prospectives of Agreement, Para 5) (Franck Report, 1945)

The Franck Report urged the Truman administration to carry out a demonstration of the atomic bomb rather than to use it without advance warning against a Japanese city. Secretary of War Stimson and his advisers, including J. Robert Oppenheimer, found the recommendations unpersuasive. The bombs were dropped on Hiroshima and Nagasaki. Szilard's worst nightmares became reality (Lanquette, 1992). In May 1946, Szilard joined with Albert Einstein, Hans A. Bethe, and other leading atomic scientists to form the Emergency Committee of Atomic Scientists. The Committee sought to further the peaceful uses of atomic energy. It carried out extensive educational programs in support of international control of the atomic bomb. But with the increase in tension between the USA and the USSR, international control of atomic weapons turned out to be impossible. Neither government ever seriously considered yielding control of its atomic weapons. In 1949, the Atomic Scientists Committee became inactive and it was officially dissolved on October 10, 1951. The USA and the USSR piled up thousands of atomic bombs and intercontinental ballistic missiles to deliver them in a situation named "mutual assured destruction" (MAD for short). This symmetry of deterrent dominated international politics for decades, until the collapse of the Soviet Union. To this day, thousands of atomic weapons remain in place and ready to fly on a few seconds' notice.

Szilard's ethical commitment never wavered. His goal became to moderate or control the atomic competition between the two great powers and prevent a nuclear war. He met with influential American and Russian politicians and scientists, gave lectures, made proposals, and wrote petitions. He succeeded once in spending an hour with Nikita Khruschev, the then leader of the Soviet Union. Eventually, however, he recognized that such individual and behind-the-scenes efforts did not succeed. To affect the policies of the US government, Szilard came to understand that it was important to engage with the political system. He proposed to establish a peace lobby that would find promising candidates for office and then supply them with the funds necessary to be nominated and elected. Szilard made this proposal in a speech to the Harvard Law School Forum and repeated the speech successfully to different college audiences. In early January 1962, he drew up a trust agreement and drafted bylaws to establish the organization that became the Council for a Livable World (CLW), with the aim of delivering "the sweet voice of reason" about nuclear weapons to Congress, the White House, and the American public (Lanquette, 1992).

This was his second creative response to an ethical dilemma. It left a lasting legacy, an organization that continues to be one of the major lobbies for peace in Washington, DC. CLW provides sophisticated technical and scientific information that helps Congress make intelligent decisions about weapons of mass destruction—nuclear, chemical, and biological weapons, nuclear non-proliferation, and other national security issues. Along with its sister organization, the Center for Arms Control and Non-Proliferation, the Council has been at the forefront of US arms control and national security policy for nearly half a century.

With the end of the Cold War in the 1990s, nuclear disarmament again became a live issue. In April 2009, President Barack Obama made a speech in Prague laying out a vision of the eventual dismantling of all nuclear weapons. In April 2010, also in Prague, Obama and Dmitiri A. Medvedev, Russia's former president, signed the Strategic Arms Reduction Treaty, an agreement to pare down both countries' nuclear arsenals. The treaty, known as New Start, was given final approval by the US Senate in December 2010. Szilard's creation, the Council for a Livable World, will continue to play an important part in decreasing the nuclear arsenal.

These events raise the issue of scientific "progress" and morality. In some cases, difficult choices are apparent from the very beginning of a major project, as was the case with the nuclear bomb. However, other circumstances present a complex mixture of solving one problem but creating others; for instance, in the overuse of antibiotics. As a society, we seem to be energized and enchanted by the relentlessness of scientific progress. It is hard to stop the momentum and create an environment for considering the moral, practical, and systemic consequences of the race for discovery. The myth of the neutrality of science, because of its use of reason and logic, interferes with the practice of reflection where ethical issues are raised and debated.

By choosing big problems and addressing them creatively, people can find themselves in situations of great moral uncertainty. The very powerful motivation that drives a physicist to understand the basic processes of the universe confronted Szilard with the consequences of his discovery, a burden from which he could never again free himself. What was exceptional about him was that, rather than escaping or denying those implications of his discovery, he devoted a good part of the rest of his life to seeking ethical and creative ways to reshape the consequences.

"Through a narrow window": Friedl Dicker-Brandeis

The story of a young art teacher illustrates how creative endeavors can be a source of solace at a time when one's life is under extreme threat. The choices that Friedl Dicker-Brandeis made when facing deportation to German concentration camps in the Second World War reveal a morality of care going beyond the struggle for individual survival. Dicker-Brandeis responded to a notification for a transport to a camp by focusing on her ability to share her creative talents. The legacy of that decision is the basis for a remarkable exhibit, curated by Professor Linney Wix of the University of New Mexico, of the paintings of Friedl Dicker-Brandeis and her students who were incarcerated in Terezín. The story of how these works were made and survived the war is one of creativity and ethics.

Friedl Dicker was born in Vienna at the end of the nineteenth century. She was an only child who lost her mother before she turned 4. Little is known of her childhood, but it has been reported that she frequently drew in the stationery shop where her father worked. The first art form that she studied was photography, in a city abundant in artistic activities and institutions. Starting at age 19, she studied painting with teachers who engaged in "reform pedagogy" (Wix, 2010, p. 8). She joined one of her instructors at the newly emerging Bauhaus program that combined art, politics, and philosophy. This influential center was situated first in Weimar, then in Dessau, and lastly in Berlin. (Once Hitler came to power, members of the Bauhaus movement spread out to many non-German communities.) Dicker studied in Weimar and was deeply influenced by the intuitive and spiritual approach of Johannes Itten and some of his colleagues during this early visionary period of Bauhaus (1919–23). Itten's notion of the student–teacher relationship was of a sharing community rather than the more usual hierarchical arrangements. He developed a key practice of the student "feeling into" the subject, form, or material, stressing the nature of the experience. He further believed that intuition and emotion were central to making art, without ignoring the more academic aspects of art education.

Friedl "emerged as a well-rounded artist with strong skills in handicrafts, a firm grasp on the foundations of art, and an ability to work intuitively and emotionally within varied art practices" (Wix, 2010, p. 12). After leaving Weimar, she turned to architectural and textile design (1923–34) and practiced these applied art forms for a decade. Then she started to paint again. Her work included flowers, landscapes, and portraits, although she never settled into a single style. "Her ability to render with exquisite sensitivity the quotidian things that brought her joy is a distinguishing feature of most of her work" (Wix, 2010, p. 13). Starting in 1931, she taught children, art students, and kindergarten teachers interested in art education. She also engaged in anti-fascist political activity, and experienced arrest and a painful interrogation. During this period she explored psychological insights about herself and the sources of her art. Her paintings included her appreciation of the physical world around her, as well as faces of friends and loved ones. In 1936, she married Pavel Brandeis in Prague. They lived in Eastern Bohemia and in 1942, they received a notice to report for deportation.

During the weeks in which she readied herself for departure, Friedl's creativity and ethical commitment joined to govern her decisions. Deportees were limited to 50 kilograms of materials to take with them. She chose to pack art materials for children, rather than cherished personal belongings, a choice that reveals her commitment to others instead of herself. It was governed by her deep belief that she and her students would find solace in artistic engagement, no matter what they would confront after deportation.

Friedl and her husband Pavel were taken to Terezín concentration camp. This differed from the other camps in that the Germans used it as an exhibit, to show to the foreign press prisoners pursuing different activities, including Friedl's art lessons. While the public face of the camp focused on such activities, everyone was aware that any day the inhabitants could be taken to an extermination camp.

Once Friedl arrived at Terezín, she started to work with the art materials she had brought with her and with additional materials she and her students could find in the camp. She gave a great deal of support to her students, encouraging them to draw what they remembered, and what they dreamed about. The approach to artistic work that she developed in Terezín allowed her and her students to live imaginatively in a very narrow and terrifying world. She achieved a delicate balance between developing their individual self-expression and fostering collaboration and community among "Friedl's girls." As she commented:

> Through the interdependency something comes about that in later life will play an enormous role; that is, the work of a group becomes not the work of competing individuals but an achievement of all of them. Through the intensive communal effort...the group and individuals achieve the best results. Under these circumstances the children also are inclined to cope with the difficulties of the sparse materials, to get by with little, to help one another, and to fit themselves into the community or group. (Dicker-Brandeis, 1943, p. 133)

Friedl helped her students be aware of their environment, including what the barracks' windows revealed. They used each other for portraits and also integrated into their paintings images from museums and art books, and scenes from their past. In some of their paintings the children presented a lost world of houses, trees, and animals—all that they had to leave behind. In addition to teaching them techniques, Friedl was encouraging them to find their own artistic expression of their many losses, transcending their captivity:

> Using art to overcome the debilitation of hunger and fear, to dispute the Nazi policy that tried to make its victims less than human, that presented them to the world and to themselves as weak and cowardly, is an act of resistance. The students' sense of aliveness to visual forms and their human bond with each other and their teacher refuted the dehumanization. (John-Steiner in Wix, 2010, p. xii)

We can only speculate about the positive impact on Friedl of being able to use her talents to work, to channel her own terrors by seeing how the children responded to her teaching and the way they developed under her care. The few children who survived recalled how the art classes helped them to deal with hunger and imprisonment, and provided opportunities to create art themselves, which their teacher shared with them. "That time when you are concentrating on something beautiful, that's how we forgot where we were" (Weissberger, 2001).

Friedl's ethical determination did not waiver, even when she was facing transportation to death at Auschwitz. She packed more than 5,000 pieces of the children's artwork into two suitcases and buried them. After the war, during which Friedl and many of the children perished, the suitcases were found and turned over to the Jewish Museum in Prague.

In addition to making art, the children and young people in Terezín performed in theater productions and wrote poetry. These became the basis of the book and theater production *I Never Saw Another Butterfly*, a depiction of the children's experience (Volavková, 1942). Accompanied by music and choreography, this moving play has made it possible for many contemporary young people to identify with the children of Terezín, most of whom perished in Auschwitz. The art exhibit, loaned from Prague, was transported to New Mexico, where many local schools brought their students to view the paintings and to share stories about the lives of these brave young people and their teacher.

Friedl's story underlines the complexity of creative endeavors. Some are born of joyous exploration, a love of the raw materials—colors, sounds, the rhythm of words, the beauty of movement—and pride in completing novel projects that are significant to one's community.

Other creative drives are the result of flight from the tragedies of life. Creative individuals and communities reach beyond a simplified image of the world:

> This is what the painter does, and the poet, the speculative philosopher, the natural scientist, each in his own way. Into this image and his formation he places the center of gravity of his emotional life in order to attain the peace and serenity that he cannot find within the narrow confines of swirling, personal experience. (Holton, 1978, pp. 231–232)

Dicker-Brandeis achieved similar "sanctuaries of sanity where making images empowered children in their struggle to survive each day" (Wix, 2010, p. 36). Teaching, even in ordinary circumstances, is an expansion of the self, reaching to the inner worlds of students. In Terezín, Friedl buttressed her own strength by transcending the threats to her life. Our next example, Clarence Stephens, was also a teacher who could identify with the struggles of his students. He, too, chose ethical solutions to human limitations and pain. But the circumstances that he confronted are more representative of how ethics and creativity can influence everyday life.

Clarence Stephens

The ethics of care in teaching mathematics was central to Clarence Stephens' passionate commitment to his students at SUNY Potsdam. They received patience, encouragement, and belief in their abilities in an environment different from the usual test-driven classroom.

In March 1987, an article appeared in the *American Mathematical Monthly* entitled "A Modern Fairy Tale." It began:

> Tucked away in a rural corner of North America lies a phenomenally successful undergraduate mathematics program. Picture a typical, publicly funded arts and science undergraduate institution of about 5,000 students, with separate departments of mathematics and computer science. While the total number of undergraduates has remained relatively fixed over the past 15 years, the number of mathematics majors has doubled and doubled again and again to over 400 now in third and fourth years. They don't offer a special curriculum. It is just a standard, traditional pure mathematics department. More than half the freshman class elects calculus, because of the reputation

of the mathematics department carried back to their high schools. And of the less than 1,000 Bachelor degrees awarded, almost 20 percent are in mathematics. In case you are unaware, one percent of bachelor degrees granted in North America are in mathematics. These students graduate with a confidence in their ability that convinces prospective employers to hire them, at I.B.M., General Dynamics, Bell Laboratories and so on.

Do they just lower their standards? Mathematics teachers in the university across the street [Clarkson Institute of Technology] say "no." They see no significant difference between their performance and that of their own students.

The students say the faculty members really care about them, care that each one can develop to the maximum possible level. It is simply the transforming power of love, love through encouragement, caring and the fostering of a supportive environment. By the time they enter the senior year, many can read and learn from mathematics texts and articles on their own. They graduate more women in mathematics than men. They redress a lack of confidence many women feel about mathematics. In the past ten years, almost every year the top graduating student at this institution, across all programs, has been a woman in mathematics. (Poland, 1987, p. 293)

The architect of this innovative approach was Clarence Stephens, a mathematics educator who encountered a contradiction familiar to every college professor of mathematics in the USA. Most college students avoid mathematics if they can, and suffer through it if they must. In the standard "pre-calculus" courses (reviewing middle-school and high-school algebra and geometry), and also in first-year calculus, retention rates are abysmal and failure rates are astonishing. Stephens had a very rare response to this unpleasant reality. He was unwilling to accept it. Instead, he challenged it, rejected the unquestioned assumptions behind it, and created a radically different college mathematics education in the 1980s, which was phenomenally successful.

In the experience of one of the authors [Hersh] and impressions from other teachers, half of the students who sign up for "developmental math" or "pre-calculus" or first-year calculus will drop out before the final exam. Of those who take the final exam, around 50 percent pass and around 50 percent fail. Overall, 25 percent success, 75 percent failure. (Of course, there is a lot of variation, depending on many factors.) These elementary math courses are prerequisites for desirable careers, not only in science and engineering, but also in medicine, business, and architecture. Many students who finally pass the course do so only after several repeated failures. Most of them would not be there at all if the course were not a prerequisite for some career goal.

This is the ethical dilemma Stephens confronted: Was he serving these students, or was he harming them, by forcing them into courses they would rather avoid, where they experienced repeated failure? Was he as a teacher permitted to participate in such a dysfunctional, destructive system of education, or was he obligated to do all in his power to change it for the better? These issues are seldom discussed, but they are central to the ethics of teaching. If participation in an educational experience diminishes the students and robs them of self-respect and confidence in approaching new challenges, then the very mission of teaching is debased. In our contemporary obsession with the "necessity" of mathematical and scientific expertise for the greatest number of students, we fail to carefully balance psychological and educational costs and advantages. One resolution of this ethical conflict is to examine the usefulness of higher mathematics as an educational filter and to rethink where, how, when, and to whom it is a necessary requirement, thus minimizing the psychological damage so many students experience. Another solution is the one followed by Clarence Stephens based on his own personal and school history.

Stephens' parents died when he was a young child, and he supported himself through high school and a bachelor's degree at Johnson C. Smith University, a traditionally black institution in North Carolina. He earned master's and doctor's degrees in mathematics at the University of Michigan. During his years as a graduate student, the University of Michigan did not employ African Americans as teaching assistants, and he earned a living by waiting on tables.

He started teaching at Prairie State University, a traditionally black institution in Texas, while still working on his PhD from Michigan. There he discovered that one of his African American students, Beauregard Stubblefield, already understood the elementary material in his course. Professor Stephens gave him a grade of "A" and made him an assistant in the course to provide a model who had the same educational experience as the other students.

In finding this special role for Stubblefield, Stephens used a creative solution, providing him with a useful role as an example to students whose fear of mathematics he did not share. (Stubblefield went on to get a PhD at Michigan, the same school Stephens had gone to, and had a long teaching career at colleges in the USA and abroad.)

After a few years at Prairie State, Stephens moved to Morgan State College, another traditionally black institution, in Baltimore. His reason for going there was to have access to the mathematics library at Johns Hopkins. He was surprised to find that Morgan State was hiring him as department chair, at the young age of 30.

Early in his teaching, Stephens had become convinced that any college student who wanted to learn college mathematics could do so, "under the right conditions." He wrote:

> When I tutored my fellow schoolmates in mathematics during my high school days, I learned that many students can learn mathematics if the learning environment is favorable. So when I was given the opportunity of leadership in the Chair of a mathematics department, I was determined to prove my conjecture. My main difficulty in proving it was that students, faculty, and administration did not believe my conjecture was true. However, in my effort to prove it, I received support from each group. The students I was trying to help gave me the best support. The results we obtained at Morgan State College and at SUNY in Potsdam proved, at least to me, that my conjecture was true. (Megginson, 2003, p. 179)

What did Stephens mean by "the right conditions"? He did not impose any particular teaching style or theory. Observers at Potsdam found that math teachers there were free to work in the style that suited them best. However, Stephens did continually implement two very non-traditional precepts.

First of all: Recognize that for most students, it takes time to assimilate new mathematical concepts. Stephens liked to say, "make haste slowly. Give the students as much time as they need to learn the material." This precept contrasts with the usual practice of maximizing the amount of material students are to learn in an emotionally neutral and frequently cold environment. Thus, his second and even more important precept: Believe in the students. Let them know you are confident that they will succeed (Megginson, 2003).

As department chair, Stephens worked slowly and patiently to win over his faculty to his educational philosophy. It directly contradicted several standard ideas prevalent in US math education: for instance, "math is only for certain people" and "the reason most people flunk is just lack of ability." Stephens' success as a teacher was indeed remarkable, but it is not unique. There are other exceptional teachers who succeed where others fail. Most remarkable, indeed unique, was his success in leading whole departments to accept and follow these beliefs.

How did he do it? With his faculty as with his students, he was gentle. He let them have all the time they needed before they could accept his unorthodox, heretical teaching philosophy. There were several essential elements to his achievement. First of all, he was able to recognize, from his own success as a high-school tutor, that students who want to learn college math can do so, under the right conditions. This meant thinking independently, based on his own observations and experience, even at the cost of disagreeing with and rejecting the received wisdom, "what everybody knows." But he went further than just thinking independently: he had the insight, the creativity, to imagine a different way of teaching college mathematics, a way to create "the right conditions." This was indeed a daring thought, for it was about much more than changing his own teaching behavior. He had the courage to undertake to change the teaching behavior of all his colleagues, many of whom certainly would at first have found his ideas to be fantastic. And not only was he able to imagine a more productive, humane way of teaching college mathematics, and able to think it possible to convert all his colleagues to such a way of teaching, he was able actually to persist, in calm, quiet, persuasive behavior year after year, in order to gradually convince and win all of them over. Stephens' manner, beliefs, patience, and knowledge all contributed to an ethical creativity that confronted current practice and presented an important model for mathematical teaching.

Conclusion

We have focused on three case studies where brave individuals drew on their creativity and fashioned ethical solutions to challenging social problems. We started with some of the most overwhelming issues scientists have confronted in the application of their discoveries. We presented an individual, Leo Szilard, who spent his life trying to prevent the use of the atomic bomb on a human population at the end of the Second World War and during the years that followed. He once argued that the nuclear crisis would not be resolved until both scientists and politicians practice "the art of the impossible": "This, I believe, might be achieved when statesmen will be more afraid of the atomic bomb than they are afraid of using their imagination, because imagination is the tool which has to be used if the impossible is to be accomplished" (Lanquette, 1992, p. 301).

A similar belief in the power of imagination fueled Friedl Dicker-Brandeis as she taught art to the young inmates of the Terezín concentration camp. For Friedl, creativity was the means for an ethical response to the tragic life that she and her students faced. She is an example of the transformation that art and teaching can accomplish even under the most inhumane circumstances. Teaching and learning are profoundly emotional as well as intellectual endeavors. Creative solutions are needed for the ethical challenges of alienation and damage to students' self-respect in many educational settings. To overcome these challenges, Clarence Stephens relied on imagination and caring to transform the status quo in mathematical education. The stories of these three individuals reveal how they used their talents and principles to change the world as they found it.

References

Csikszentmihalyi, M. (1990). *Flow: The Psychology of Optimal Experience*. New York: HarperCollins.

Dicker-Brandeis, F. (1943). On Children's Art (L. Foster, trans., 2005). Unpublished manuscript.

Franck Report. (1945). Report of the Committee on Political and Social Problems Manhattan Project "Metallurgical Laboratory" University of Chicago, June 11, 1945. US National Archives, Washington DC: Record Group 77, Manhattan Engineer District Records, Harrison-Bundy File, folder #76. http://www.dannen.com/decision/franck.html, accessed 10 November 2013.

Holton, G. (1978). *The Scientific Imagination: Case Studies*. Cambridge: Cambridge University Press.

John-Steiner, V. (1985). *Notebooks of the Mind*. Albuquerque, NM: University of New Mexico Press.

Lanquette, W. (1992). *A Genius in the Shadows: A Biography of Leo Szilard*. New York: Scribner.

Megginson, R. E. (2003). Yueh-Gin Gung and Dr Charles Y. Hu award to Clarence F. Stephens for distinguished service to mathematics. *American Mathematical Monthly 110*(3), 177–180.

Poland, J. (1987). A modern fairy tale. *American Mathematical Monthly, 94*(3), 293.

Volavková, H. (ed.) (1993). *I Never Saw Another Butterfly: Children's Drawings and Poems from Terezín Concentration Camp, 1942–1944*. 2nd edn by the United States Holocaust Museum. New York: Schocken Books.

Wallace, D. B., & Gruber, H. E. (eds.) (1989). *Creative People at Work*. New York: Oxford University Press.

Weissberger, E. (2001). Personal interview with Linny Wix, July 24.

Wix, L. (2010). *Through a Narrow Window: Friedl Dicker-Brandeis and her Terezín Students*. Albuquerque, NM: University of New Mexico Press.

12

The Hacker Ethic for Gifted Scientists

Kirsi Tirri
University of Helsinki, Finland

This chapter explores the ethics of gifted people in science with a special emphasis on a "hacker ethic." Gifted students and professionals in science create new ideas and products that can be used in the benefit of our society. However, the creative process in science includes many ethical issues that need to be considered before publishing the new idea or the product. Combining excellence with ethics relates to ethical models developed in the academic context, such as Pekka Himanen's theoretical approach to the hacker ethic. In his work, Himanen (2001) introduced a new kind of ethic, the "hacker work ethic," which has replaced the dominance of the Protestant work ethic with a passionate attitude and relationship to one's work. With the word "hackers," he referred to people who did their work because of intrinsic interest, excitement, and joy, whereas the Protestant work ethic emphasized work as a duty and a calling. The successful scientists resemble the hackers with their strong inner drive to excel (Koro-Ljungberg & Tirri, 2002; Tirri & Campbell, 2002).

Hackers wanted to realize their passion together with others, and they wanted to create something valuable for the community and be recognized for that by their peers. A passionate attitude toward work, a desire to learn more about subjects and phenomena, was not an attitude found only among computer hackers in Himanen's study, but also among science researchers. Himanen (2001) identified the monastery as the historical precursor to the Protestant ethic; for the hacker work ethic, the academic community was the precursor. Historically, the academic community has always defended a person's freedom to organize his or her work schedule. Hackers did not organize their lives in terms of routines and continuously optimized workdays, but in terms of a dynamic flow between creative work and life's other passions. Scientists working within the academy were occasionally free to organize their days according to a flow between creative work and other passions, but equally often, scientists were committed to rigid schedules, timelines, and assigned tasks (Koro-Ljungberg & Tirri, 2002).

According to Himanen (2001), the hacker work ethic consisted of "melding passion with freedom." For the hackers, recognition within a community that shares their passion was more important and more deeply satisfying than money, just as it was for scholars in the academy. However, this recognition was no substitute for passion—it had to come as a result of passionate action, of the creation of something socially valuable to this creative community (Himanen, 2001). A "hacker" who lived according to the hacker ethic on all three of these levels—work, money, ethic—gained the highest respect in his or her community (Himanen, 2001).

In this chapter, case studies of gifted science students and researchers are presented with the emphasis on how to combine ethics with creativity. First, ethical sensitivity is described as an important skill in identifying new ethical dilemmas and finding new ways to discuss and negotiate creative solutions to them. Second, three qualitative case studies on science and ethics are introduced; and last, the implications of these studies in educating creative and moral people in science are discussed.

What is ethical sensitivity?

Skills in moral judgment and especially in moral sensitivity are necessary in order to combine excellence with ethics (Tirri, 2011). In this chapter ethical sensitivity is also seen as a necessary skill in combining creativity with ethics. High-ability students have been shown to be superior in moral judgment when compared to students of average ability. However, high academic ability does not always predict high moral judgment (Narvaez, 1993; Tirri, 2011). Morality includes other components as well, such as sensitivity, motivation, and character. In the research on the ethical sensitivity of gifted people in science, we have used the definition of moral sensitivity by Bebeau, Rest, and Narvaez (1999). According to them, moral sensitivity is about the awareness of how our actions affect other people. Thus, without possessing a moral sensitivity it would be difficult to see the kind of moral issues that are involved in science. However, to respond to a situation in a moral way, a scientist must be able to perceive and interpret events in a way that leads to ethical action. A morally sensitive scientist notes various situational cues and is able to visualize several alternative actions in response to that situation. He or she draws on many aspects, skills, techniques, and components of interpersonal sensitivity. These include taking the perspective of others (role-taking), cultivating empathy for others, and interpreting a situation based on imagining what might happen and who might be affected. Moral sensitivity is closely related to a new suggested intelligence type, social intelligence, which can be defined as the ability to get along well with others and get them to cooperate with you (Albrecht, 2006; Goleman, 2006).

Ethical sensitivity includes similar components to hacker ethics. Hackers wanted to realize their passion together with others, and they wanted to create something valuable for the community and be recognized for that by their peers. In a similar way, ethical sensitivity builds on caring and communication with the idea of finding new, innovative solutions to ethical dilemmas in the community of ethically sensitive people.

Empirical studies on ethical sensitivity

Numerous tests of ethical sensitivity have been developed over the years, but most of them are very context specific; for example, relating to medicine and dental education (Bebeau, Rest & Yamoor, 1985) or to racial and gender intolerance (Brabeck et al., 2000). We have developed an instrument to study ethical sensitivity in educational contexts with diverse populations (Tirri & Nokelainen, 2007, 2011). The Ethical Sensitivity Scale Questionnaire (ESSQ) was operationalized based on the theory by Narvaez (2001). In that theory, ethical sensitivity includes the following seven dimensions: (1) reading and expressing emotions; (2) taking the perspectives of others; (3) caring by connecting to others; (4) working with interpersonal and group differences; (5) preventing social bias; (6) generating interpretations and options; and (7) identifying the consequences of actions and options.

The instrument consists of 28 Likert-scale items with the response options from 1 (totally disagree) to 5 (totally agree). The ESSQ items are designed in such a way that they apply to people from different backgrounds and cultures. This allows us to use the instrument in a multicultural society and in cross-cultural studies. The statements describe the issues and values that the respondent finds important. Each of the seven dimensions is operationalized in the questionnaire with four statements (Tirri & Nokelainen, 2007, 2011).

The instrument has been used in different cultural contexts with Finnish, Iranian, and Dutch teachers and students (Gholami & Tirri, 2012a; Kuusisto, Tirri, & Rissanen, 2012; Schutte, Tirri, & Wolfersberger, under review). Our results indicate that the instrument is reliable across different cultures and yields similar trends with different populations. Generally, teachers evaluate their ethical sensitivity as quite high with an emphasis on caring ethics. However, in a recent study on teachers' ethical sensitivity, science teachers rated their ethical sensitivity lower than teachers of other subjects (Kuusisto, Tirri, & Rissanen, 2012). In other studies, females and high-ability students tend to score higher in ethical sensitivity than males or average-ability students (Gholami & Tirri, 2012a, b; Schutte, Wolfersberger & Tirri, under review; Tirri & Nokelainen, 2007). This finding supports other researchers' notion that gifted students held a privileged position in the maturation of moral thinking because of their precocious intellectual growth (Andreani & Pagnin, 1993; Karnes & Brown, 1981; Terman, 1925). The most relevant of these empirical findings for this chapter is the tendency of science teachers to score lower in ethical sensitivity than teachers of other subjects. In the following case studies the ethical sensitivity of gifted high school students and researchers are explored in more detail.

Qualitative studies on ethical sensitivity

Study 1. Gifted high school science students

In a study on the moral reasoning and scientific argumentation of Finnish adolescents (N = 31) who are gifted in science (Tirri & Pehkonen, 2002), attention was also paid to their moral sensitivity. In qualitative essays and interviews, the students were asked to identify moral dilemmas in science and to provide solutions to them. The findings show that the students identify different aspects as relevant in discussing the same moral dilemma. Furthermore, the principles and values used in solving the dilemma reveal qualitative differences in students' moral sensitivity. Students' arguments and justifications for doing archeological research in graves were analyzed with the help of technical terminology developed by Toulmin (1958). All the students reached the conclusion that research in graves is morally justified. However, those students who gained the highest scores in the Defining Issues Test (DIT) reflected on the dilemma with different justifications and values than the students who attained only average scores. Furthermore, girls and boys differed qualitatively from each other in their argumentation process (Tirri & Pehkonen, 2002).

Two illustrative cases reflected some general trends in the argumentation pattern of the whole group studied (Tirri & Pehkonen, 2002). Tina is a good example of a gifted girl whose argumentation is logical and elegant and who provides theoretical and ethical backing for her arguments. Furthermore, she demonstrates emotional and spiritual sensitivity in her reflection on moral dilemmas in science. Tina's argument for the disadvantages of archeological research in graves is based on her moral sensitivity in respecting other people's feelings and values. She does not limit her arguments to rational and scientific evidence, considering the sacred nature of the grave and understanding the religious concept of holiness. The backing she uses is based on ethical values of respecting things that are considered holy by some people. Tina also demonstrates very good judgment in her reflection on the advantages of archeological studies. She admits that sometimes the graves have to be studied; however, the reasons need to be valid and the grave must be old enough. For her final conclusion, Tina states that we should negotiate contracts and laws on how to conduct archeological studies in graves. Compared to the whole group of gifted students, her argument is outstanding, including critical thinking, logic, and moral sensitivity. Her final conclusion provides some concrete but not naïve ways to approach this dilemma. It is no surprise that Tina ranked at the highest level of post-conventional moral reasoning in the DIT.

Alex's argumentation for the advantages of archeological research in graves was typical of the whole group of gifted students. The majority of the arguments students used to justify conduct in science were based on utilitarian ethics. The new knowledge in science was identified as the leading value that brings the greatest benefit to people. However, in most cases the students acknowledged the need to provide some exceptions to this rule. Even in science, the researchers should pay attention to the feelings of those people who are affected by the research. In this study, archeological research in graves was advocated, with the exception of those graves that are so new that the relatives of the dead could still be alive. Alex's argument for the disadvantages of archeological research in graves was not typical of the group of gifted students as a whole, however. His strong identification with the scientists led him to consider scientific moral dilemmas only from the scientists' point of view. This emphasis made him neglect the relatives and other people who may have different values concerning studies in graves. Alex's score in the DIT was below average compared to the whole group of gifted students. Compared to his almost perfect score in the Raven test, the authors claim that his general intellectual ability is more developed than his moral reasoning. The qualitative study of his argumentation may reveal in part the reasons for his only average score in the DIT. Alex's thinking lacks universal principles in his reflection on the disadvantages of doing research in graves. He considers the dilemma solely from his own, partial standpoint. However, the ability of an impartial moral agent to use universal moral judgments is considered to be the most mature moral reasoning, according to Kohlberg's procedures (Strike, 1999).

The results of the study reveal that there are qualitative differences in the moral reasoning of gifted adolescents. High intellectual ability does not predict mature moral judgment. Furthermore, responsible moral judgments regarding the moral dilemmas in science require moral motivation and moral sensitivity. Teachers and educators should nurture the moral growth of future scientists by exploring and discussing the ethical aspects of doing scientific studies. The argumentation model presented in the article can serve as a pedagogical tool for teachers in reflecting on moral dilemmas in science with their students (Tirri & Pehkonen, 2002).

Study 2. Gifted international science students

In the second study, international high school students from Europe and Asia who were gifted in science asked scientific, societal, and moral questions related to the themes in science they wanted to study (Tirri et al., 2012). The students were identified as gifted because most of them came from scientific schools or classes, and many had done well in national or international science competitions, won scholarships and prizes, as well as had good school grades. An equal amount of males and females applied to the Millennium Youth Camp, but males applied more to the information technology and math group, which was not considered in this study. Females applied more to the climate change, water, and renewable resources groups. Females' interest in these themes can be explained by the Relevance of Science Research (ROSE; Sjøberg & Schreiner, 2010), where researchers conclude that males, more often than females, think that threats to the environment are not their personal business. Females also tend to believe that they can personally influence what happens with the environment. Our findings demonstrate that the most often asked type of question was scientific in nature. The boys asked more scientific questions than the girls. This characteristic was also evident in our earlier studies (Tirri, Tallent-Runnels, & Nokelainen, 2005; Tirri & Nokelainen, 2006). The girls asked more societal questions than the boys. Both boys and girls asked the same number of moral questions related to their interests. Moral questions were not as common as the other question types in our study. This feature was also evident in our earlier studies with gifted preadolescents and adolescents. In those studies we demonstrated that there tend to be fewer moral questions as students get older (Tirri & Nokelainen, 2006). The students in the climate change group asked more moral questions than their peers in the water group. The nature of the moral questions asked by the girls was more personal than the moral questions asked by the boys. This could be explained by the ethics of care that are so typical of girls and women, who tend to care, protect, and show empathy to others in more personal ways than men (Tirri, 2003). The boys took a more impersonal approach to moral questions in science, but their questions often involved a future perspective. Moral questions in science need to be discussed and solved with a view to the future of humankind. Many moral issues related to climate change, renewable energy, and water require solutions that have an understanding of future needs. We need to protect, restore, and nurture our environment to be able to have a future on earth.

In science education, teachers should actively discuss the moral questions that science raises with both a caring and a long-term perspective. One way to increase the focus on morality and ethical issues in a science curriculum is to bring more socio-scientific issues into education. Another means is to teach on the Nature of Science (NOS). Through this, students gain understanding of how science advances and what kind of decisions scientists need to make in their careers. Furthermore, students see themselves as decision-makers, possibly increasing their interest in considering moral issues. The findings of this study were used to improve the next camp by taking moral issues more into consideration in the project work done by the young people. An example of this is the project work of the renewable energy group, where students had to consider, among other things, the use of food crops as a source of fuel. Gifted students in science may have the best cognitive skills and logical thinking, but they may lack the ethical sensitivity that is needed to solve moral dilemmas in science (Tirri & Pehkonen, 2002). This study points to the need for teachers to teach socio-scientific issues and discuss moral questions in science, which might influence the future of humankind. Science teaching has a moral core like all the other subjects taught in high schools all over the world.

Study 3. Gifted researchers in science

In a study of adult gifted females, including Academic Olympians, the importance of beliefs and values in their lives was established

(Tirri, 2002; Tirri & Koro-Ljungberg, 2002). In another study including researchers, the beliefs and values that guide the academic work of
successful Finnish scientists (N = 16) were identified (Koro-Ljungberg &
Tirri, 2002). In both studies interviews were conducted in which the
professional and personal lives of these individuals were discussed, as
well as themes related to their choices of career, job, spouse, lifestyle,
friends, and hobbies. All the scientists in the study worked in academic
research environments (Tirri, 2001). One of the aims was to reveal the
contextual and situational nature of the academic work ethic. Before this
study, other researchers acknowledged the situational and contextual
essence of Kohlberg's (1969) and Gilligan's (1982) moral orientations.
Strike (1999) has argued that justice and caring aim at different moral
goods. According to Noddings (1999), the ethical orientations of justice
and care might also apply to the different domains. In some contexts, justice and care might work together to produce a genuinely
moral solution, and on other occasions they might conflict with one
another.

When placing ethical orientations of justice and care in the context
of scientific work, it is often assumed that scholars need to acknowledge
both forms of ethics in order to produce good science and build a just
and caring society through their work. In academia, some general rules
to evaluate the research findings and accomplishments of the members
needs to be established. Every scholar wants to be treated fairly and
have equal opportunities to publish his or her work and compete for
academic rankings. In addition to equal rights, a good scientific community must acknowledge individual differences and build a sense of
belonging among its members (Koro-Ljungberg & Tirri, 2002).

In the study by Koro-Ljungberg and Tirri (2002), it was argued that in
the context of academia justice and care do not rule separate spheres,
nor does one dominate the other. Scientists' narratives revealed that
various ethical orientations sometimes work together and sometimes
conflict with one another, as suggested by Strike (1999). Furthermore,
the authors found that the ethical orientations of care and justice were
insufficient to explain the essence of the moral orientations among
the scientists studied. Therefore, they propose that the conceptualizations and understandings of scientists' work ethics must go beyond
justice and care-oriented reasoning. Instead, ethical analyses and moral
perceptions should create new horizons and give inner meanings to
both scientists themselves and to their communities. Based on the ethical analyses, the concepts of ethics of justice, care, and empowerment
were introduced as possible value systems guiding researchers' work

ethics. After the discovery of the theoretical misfit between the data and Kohlberg and Gilligan's models, as well as the lack of emphasis within the empowering self as a source of moral arguments and actions, a concept was developed of an ethic of empowerment to describe values and beliefs connected to the moral practices of enabled, situated selves. The ethic of empowerment describes the values and beliefs related to academic motivation, self-image, and the academic work culture. It is practiced through multiple conflicting subjectivities, which allow scientists to follow their internal voices, and to change and reevaluate their ethical assumptions, values, and beliefs related to their personal and professional lives (Koro-Ljungberg & Tirri, 2002).

Conclusion

In this chapter a hacker ethic is introduced as a new approach to combine ethics with creativity in science. Ethical sensitivity is identified as a necessary skill for scientists to identify and solve ethical dilemmas in caring and communicative ways. Three case studies on gifted science students and researchers are presented to demonstrate the nature of ethics needed in scientific studies and work. In the first case study with high school students concerning the moral dilemma of archeological studies in graves, analysis of the argument demonstrates that responsible moral judgments for the moral dilemmas in science require moral motivation and moral sensitivity. In this dilemma, ethically sensitive and creative solutions are needed together with communication and negotiation skills. The teachers can use the model of argument presented in the study to discuss the moral dilemmas in science with their students. Ethically sensitive and creative solutions should be encouraged and modeled in science teaching.

The second case study explores further the questions of gifted international high school students. This study points to the need for teachers to teach socio-scientific issues as part of the science curriculum and to discuss moral questions in science, which might influence the future of humankind. The Millennium Youth Camp provides a great opportunity for gifted international high school students to meet like-minded friends and to be challenged both academically and socially. Furthermore, this kind of international summer camp has a strong emphasis on global responsibility. Therefore, it covers many aspects of social, emotional, and moral education that have been neglected in gifted and science education. It also provides the peers that are needed in the hacker ethic to inspire students to give their best and passionate effort in their studies.

The third study provides more evidence that ethical orientations of care and justice are insufficient to explain the essence of the moral orientation among the scientists. Therefore, the conceptualizations and understandings of scientists' work ethics must go beyond justice and care-oriented reasoning. Gifted, creative scientists need an ethics of empowerment that is built on their own inner drive to excel and create new things. The hacker work ethic includes many aspects that are suited to gifted and creative minds and thus helps scientists to combine ethics with creativity.

References

Albrecht, K. (2006). *Social Intelligence: The New Science of Success.* San Francisco, CA: Jossey-Bass.

Andreani, O., & Pagnin, A. (1993). Nurturing the moral development of the gifted. In K. Heller, F. Mönks, & H. Passow (eds.), *International Handbook of Research and Development of Giftedness and Talent* (pp. 539–553). Oxford: Pergamon Press.

Bebeau, M., Rest, J., & Narvaez, D. (1999). Beyond the promise: A perspective on research in moral education. *Educational Researcher, 28*(4), 18–26.

Bebeau, M., Rest, J., & Yamoor, C. (1985). Measuring dental students' ethical sensitivity. *Journal of Dental Education, 49*(4), 225–235.

Brabeck, M., Rogers, L., Sirin, S., Handerson, J., Ting, K., & Benvenuto, M. (2000). Increasing ethical sensitivity to racial and gender intolerance in schools: Development of the racial ethical sensitivity test (REST). *Ethics and Behavior, 10*(2), 119–137.

Gholami, K., & Tirri, K. (2012a). The cultural dependence of the ethical sensitivity scale questionnaire: The case of Iranian Kurdish teachers. *Education Research International, 2012.*

Gholami, K., & Tirri, K. (2012b). The teachers' perceived dimensions of caring practice: A quantitative reflection on the moral aspect of teaching. *Education Research International, 2012.*

Gilligan, C. (1982). *In a Different Voice.* Cambridge, MA: Harvard University Press.

Goleman, D. (2006). *Social Intelligence.* New York: Bantam Books.

Himanen, P. (2001). *The Hacker Ethic and the Spirit of the Information Age.* London: Vintage.

Karnes, F., & Brown, K. (1981). Moral development and the gifted: An initial investigation. *Roeper Review, 3,* 8–10.

Kohlberg, L. (1969). Stage and sequence: The cognitive-developmental approach to socialization. In D. A. Goslin (ed.), *Handbook of Socialization Theory and Research* (pp. 347–480). Chicago, IL: Rand McNally.

Koro-Ljungberg, M., & Tirri, K. (2002). Beliefs and values of successful scientists. *Journal of Beliefs and Values, 23*(2), 141–155.

Kuusisto, E., Tirri, K., & Rissanen, I. (2012). Finnish teachers' ethical sensitivity. *Education Research International, 2012.*

Narvaez, D. (1993). High achieving students and moral judgment. *Journal for the Education of the Gifted, 16*(3), 268–279.

Narvaez, D. (2001). Ethical sensitivity: Activity booklet 1. http://www.nd.edu/ ~dnarvaez/, accessed March 2, 2007.

Noddings, N. (1999). Introduction. In M. Katz, N. Noddings, & K. Strike (eds.), *Justice and Caring: The Search for Common Ground in Education* (pp. 21–37). New York: Teachers College Press.

Schutte, I., Wolfersberger, M., & Tirri, K. (under review). The relationship between ethical sensitivity, high ability and gender in higher education students.

Sjøberg, S., & Schreiner, C. (2010). The ROSE Project. An Overview and Key Findings, Oslo: University of Oslo.

Strike, K. (1999). Justice, caring, and universality: In defence of moral pluralism. In M. Katz, N. Noddings, & K. Strike (eds.), *Justice and Caring: The Search for Common Ground in Education* (pp. 21–37). New York: Teachers College Press.

Terman, L. (1925). *Genetic Studies of Genius: Vol. 1. Mental and Physical Traits of a Thousand Gifted Children.* Stanford, CA: Stanford University Press.

Tirri, K. (2001). Finland Olympiad studies: What factors contribute to the development of academic talent in Finland? *Educating Able Children, 5*(2), 56–66.

Tirri, K. (2002). Developing females' talent: Case studies of Finnish Olympians. *Journal of Research in Education, 12*(1), 80–85.

Tirri, K. (2003). The moral concerns and orientations of sixth- and ninth-grade students, *Educational Research and Evaluation, 9*(1), pp. 93–108.

Tirri, K. (2011). Combining excellence and ethics: Implications for moral education for the gifted. *Roeper Review, 33*(1), 59–64.

Tirri, K., & Campbell, J. R. (2002). Actualizing mathematical giftedness in adulthood. *Educating Able Children, 6*(1), 14–20.

Tirri, K., & Koro-Ljungberg, M. (2002). Critical incidents in the lives of gifted female Finnish scientists. *Journal of Secondary Gifted Education, 13*(4), 151–162.

Tirri, K., &. Nokelainen, P. (2006). Gifted students and the future. *KEDI Journal of Educational Policy, 3*(2), 55–66,

Tirri, K., & Nokelainen, P. (2007). Comparison of academically average and gifted students' self-rated ethical sensitivity. *Educational Research and Evaluation, 13*(6), 587–601.

Tirri, K., & Nokelainen, P. (2011). *Measuring Multiple Intelligences and Moral Sensitivities in Education.* Rotterdam: Sense Publishers.

Tirri, K., & Pehkonen, L. (2002). The moral reasoning and scientific argumentation of gifted adolescents. *Journal of Secondary Gifted Education, 13*(3), 120–129.

Tirri, K., Tallent-Runnels, M. &. Nokelainen, P. (2005). A cross-cultural study of pre-adolescents' moral, religious and spiritual questions. *British Journal of Religious Education, 27*(3), 207–214.

Tirri, K., Tolppanen, S., Aksela, M., & Kuusisto, E. (2012). A cross-cultural study of gifted students' scientific, societal, and moral questions concerning science. *Educational Research International, 2012*, Article ID 673645, doi:10.1155/2012/673645.

Toulmin, S. (1958). *The Uses of Argument.* New York: Cambridge University Press.

13

The Dialogic Witness: New Metaphors of Creative and Ethical Work in Documentary Photography

Charlotte Dixon and Helen Haste
Harvard Graduate School of Education, USA

Howard Gardner described the troubling relationship between creativity and ethics in the sciences: "On the one hand, science and innovation proceed apace, ever conquering new frontiers. On the other hand, traditional restraints against wanton experimentation and abuse appear to be tenuous" (2006, p. 53). In domains like the sciences that have well-defined professional structures, an effective approach to issues at the intersection of creativity, ethics, and morality may be an appeal, like Gardner's, to practitioners and educators working in the field. The domain of photography suffers from a similar imbalance between the application of innovative technology and attention to ethical or moral concerns. But characteristics of the field, such as porous boundaries between genres, a loose professional structure, and limited gatekeeping capacity, mitigate the effectiveness of addressing issues from within the professional realm. Innovations in the domain have enabled all of us to make, produce, and publish photographs, blurring the line between amateur and professional, making matters of ethical and creative significance both harder to address and of greater concern. Documentary photography, a genre that includes a variety of styles, forms, and purposes defined by focused subject matter, not aesthetic form (Abbot, 2010; Bogre, 2012), is of particular relevance to these issues. This chapter takes a broad and historical view of the domain and identifies culturally embedded assumptions that threaten creative efforts and undermine ethical practices.

Innovation proceeds apace, restraints are tenuous

Many people reading this chapter will have lived through the radical shift from analog to digital technology for photographic capture, processing, production, and dissemination. In 30 years, innovations in technology have made obsolete rolls of film, lengthy processing times, and expensive printing, doing away with many specialized roles within the field. The most dramatic transformation is arguably the integration of imaging and communication technologies in mobile devices. These tools afford the global "citizenry of photography" (Azoulay, 2012) with a cheap and easy way to produce images that can be instantly transmitted to individuals, web platforms, and the media. The widespread adoption of camera phones for both amateur and professional use has changed how and why visual information is created, used, shared, and evaluated by both the press and public citizens (Foster, 2009; Moller, 2012; Ritchin, 2010, 2012; Wolf & Hohfeld, 2012; Zelizer, 2010).

Access to affordable tools for image-making democratizes the field, giving individuals and social groups of all classifications a new way to create and contribute to stories of personal and civic importance. Marginalized groups, communities, and cultures, frequently the photographic subjects of an outsider, are given agency to construct and publish their own visual narratives. These new capacities generate creative potential, but also they increase the scale and scope of possible unethical or immoral treatment, such as privacy violations, misrepresentation, exploitation, and humiliation. Seasoned press photographers worry that objectionable practices will gradually erode the moral value and power of "legitimate" documentary images (Abbott, 2010). This complaint, like many others, has confronted the medium throughout its history, but some are new: the vast number of publicly accessed venues cannot be effectively regulated; gatekeepers have insufficient capacity and limited authority to monitor what is ethically acceptable or creatively admissible (Bogre, 2012; Klein, 2010); and images posted online circulate unchecked in an indeterminate afterlife as raw material for new purposes (see Batchen et al., 2012; Harriman & Lucaites, 2007; Ritchin, 2010, 2012).

Central to these concerns is the use of camera phones to transmit images immediately, without sufficient time to reflect on the consequences, in situations that are burdened with complex ethical or moral issues. For example, in 2007, New York City Mayor Michael Bloomberg called on citizens to help fight crime by using the camera phone to transmit images of suspicious activity to emergency dispatchers. Understanding the implications of the request requires looking at what was being taken for granted about photography. Things that are taken for granted presume a common cultural meaning that does not need to be justified (Billig, 1995; Haste, 2008, 2013). Identifying camera phones as tools for crime fighting belies an assumed link between an objective reality and photography, an association that goes back to the invention of the medium. This discourse frames the camera as a "truth telling device" (Sekula, 2006; see also Aker, 2012; Newton, 2001), the image as the evidential truth of an objective reality and the photographer's role to "bear witness" (Sontag, 2003, p. 56; also Chapnick, 1994; Light, 2000). Photography's relationship to witnessing "the truth" has endured and has morphed into today's "mobile witness" (Reading, 2009).

The "realness" of the image is a culturally assumed lay theory that has persisted in the face of gross advances in photographic technology, the democratization of the field, multiple divisions of the domain, and culturally systemic epistemological shifts. Postmodern and poststructuralist photography critics vehemently and unequivocally attacked the "authenticity" and "realness" of photographs for decades (for example, Barthes, 1981; Rosler, 1992; Sekula, 1982; Sontag, 1966, 1990, 2003), but Realist and Modernist associations with the truth and "what's really there" (Bogre, 2012; Linfield, 2010) linger in assumptions about photography. Thanks to public debates about faked or manipulated images in the media, most people know that a person can be made to look like a Smurf with the filters of an imaging application. The common perception is that manipulation takes place after the image is taken, while the image captured by the camera depicts an external reality (Foster, 2009; Klein, 2010; Ritchin, 2010). Debates over digital fakery underscore a belief that the controversy resides in what happens after the photograph is taken. The *original* photographic image is perceived as "real" (Moller, 2012; Ritchin, 2010), not constructed through the creative decisions of a photographer engaged in a highly contextualized social activity and mediated through a tool that optically distorts, compresses, or expands spatial relationships, and isolates moments from a single vantage point. To understand why this is the case, it is necessary to look at metaphors woven through the domain's history, how those metaphors gained a stronghold, and the reasons they are maintained against current epistemological beliefs, increased access to photographic tools, and greater sophistication around visual culture.

Embedded metaphors: Truth and the mechanical eye

Cultural psychologists have theorized the framing power of metaphors and their ability to shape popular assumptions and understandings (Haste, 1993, 2011). Metaphors are cultural constructions that allow people to better comprehend and more easily remember new concepts, but may also entrap understanding in readily available, entrenched, or stereotypical modes of thought (Haste, 1994, 2013; Keats, 2010; Lakoff & Johnson, 2003). Common metaphors like "a roadmap to peace" or "political fallout" are catchy, easy to remember, and provide visual clues that dramatize meaning. They also shape, consciously or not, the way we conceptualize those meanings. Peace, for example, would be approached differently with the working metaphor "a Venn diagram of peace" instead of "a road map."

Contemporary metaphorical descriptions of what photography is, how it works, and who makes it have roots in the first decades of the medium (Morris, 2011; Ritchin, 2010; Tremain, 2000). Pioneering inventor William Henry Fox Talbot worked in conjunction with the chemist and astronomer John Herschel to develop the calotype, a negative to positive process that most directly informed the development of film later in the nineteenth century. Talbot labeled the process "the pencil of nature" to emphasize its inherent connection to the natural world (Schaaf, 2000) and to suggest that "in place of an individual artist possessed of certain talents and aptitudes, nature now inscribes itself by itself" (Azoulay, 2012, p. 11). Talbot's conception of photography generated controversy and opposing lines of discourse eventually consolidated around two different interpretations of the medium: a discourse aligning it with mechanical objectivity and Realism; and a discourse aligning it with creative subjectivities and art. Issues of importance to the documentary tradition of photography can be traced back to the influence of Talbot and the alignment of photography with mechanical objectivity.

The invention of photography took place through both isolated and combined efforts of scientists in England and France during a period of cultural shift from Romanticism in favor of Realism, an epistemology that lent support and credibility to the development and appreciation of photography. The perceived verisimilitude of the image to the object photographed was reinforced by the growing cultural interest in Realism and was equated with the realization of a visible truth (Thompson, 2003). Complementing the connection to realism and truth, descriptions of photography at the time emphasized that photographs were made by the camera, a mechanical device, and thus distanced from the hand and consequently the influence of a human. The alignment of photography with science also supports a belief in photographic objectivity, a truth undistorted by the creative efforts of a photographer. Civil War photographer Matthew Brady coined a new metaphor for the medium, "the camera is the eye of history" (quoted in Sontag, 2003, p. 53), and his photographic assistant, Alexander Gardner, suggested that their Civil War images *witnessed* the horrific truth of war and that the fact of those images should end all wars (Sontag, 2003; Stern, 2012; Trachtenberg, 1990). Although the term "documentary" was not applied to photography until several decades later, Civil War photographs were presented and written about as evidential documents of the "truth" of war. Gardner connected the medium to the moral imperative of ending war while at the same time reinforcing the perception that the *fact* of the images, not the work of the photographer, was responsible for the persuasiveness of the message (Freund, 1982; Schiller, 1977). Indeed, "the eye of history" was not the photographer but the camera itself.

The passive role of the photographer was embedded in the early language around photography and popularized with the introduction of Kodak's Brownie camera. Kodak claimed that all the skills necessary to use a Brownie could be learned in less than 30 minutes by any child or adult. The camera's catchy advertising campaign, "You press the button, we do the rest," fueled its commercial success (West, 2000). Today we see this legacy in the language used to describe the photographic process: we point and shoot. We also commonly say that photographs are taken or captured as opposed to made or created, reinforcing a belief that cameras record the subject of an existing reality, again dismissing the role of the photographer. A clear statement of this can be found in Howard Chapnick's seminal and still popular text for students of photojournalism: "Photography, as witness to history, gives testimony in the court of public opinion. Photojournalists are the bearers of that witness" (1994, p. 13). The emphasis on mechanical objectivity, the framing of the photographer as subservient to both the camera and the subject, girded the impression that there was an objective reality to record and neutralized the issue of individual subjectivity (Chapnick, 1994; Sekula, 2006).

Although today there is greater cultural sophistication around subjectivities and multiple truths, writings and exhibitions continue to fuel entrenched metaphors that frame the photographer as an objective witness who visually records an external truth (Heifferman, 2012; Moller, 2012; Ritchin, 2010). Vilem Flusser's influential *Towards a Philosophy of Photography* posits that photographic acts are entirely "programmed" by the rigidity of the camera, and the photographer lacks all agency unless the "automated apparatus" is outwitted (2000, pp. 76–81). Discussions of truth and objective reality appear frequently in writings on photography (for example, Heifferman, 2012; Light, 2000; Linfield, 2010; Thompson, 2003; Zelizer, 2010). A recent Getty Museum exhibition entitled *Engaged Observers: Documentary Photography since the Sixties* features photographers "[e]mbracing the gray area between objectivity and subjectivity, information and interpretation, journalism and art" (Abbott, 2010, p. 1). The exhibition challenges the metaphor of the disengaged or objective witness, but the photographer is portrayed as an observer, neither objective nor subjective, thus conjuring up the Realist and Modernist principles behind those metaphors.

While studies in journalism have adopted postmodern perspectives that shift away from objectivity and toward stances that include reader/viewer agency (Soffer, 2009), the field of photography lags with only slow movement toward a more complex and contemporary understanding of truth, accuracy, and reality. The photograph is still treated as documented, embodied meaning that expresses one truth, a monolithic perspective that excludes viewer agency. Photography theorists express hope that mass exposure to digital photography as a consumer and creator has "planted an awareness in the public's mind that photographs are interpretations and constructions of reality and not parts of it" (Aker, 2012, p. 336). However, more needs to be done to break down the link between objectivity and the camera, embrace the creative process of constructing visual narratives, and engage the viewer to interact with those narratives not as objective truths but as products of complex cultural dialogues.

Moral creativity and social documentarians

Photographers, both amateur and professional, have used photographs to fight social problems and moral injustice since the Civil War. It was assumed that since photographs depicted the truth of a situation, people would be called into action if they saw photographic evidence of social injustices. This relationship between seeing a photograph and being moved to action has been significantly debated for over 100 years. Sontag (1966, 1990) and other postmodern theorists (Rosler, 1992; Sekula, 1982) insist that photographs inspire apathy not action. Critics of those postmodern theorists (for example, Bogre, 2012; Linfield, 2010) point to numerous examples spanning the history of photography that demonstrate the effectiveness of images to generate change. Some photographers, such as Jacob Riis, Lewis Hine, Eugene Smith, Sebastiaño Salgado, and others, produced series that have an enduring legacy in social reform. Others, most notably social documentary photographers practicing from the 1930s through to the 1970s, successfully altered social attitudes. The work of these photographers illustrates what has been described as moral creativity (Haste, 1993). The three components of moral creativity—vision (to see what others have conventionalized or failed to identify), efficacy (understanding that one can act on issues of importance), and responsibility (the outcomes of acting on one's vision)—are well illustrated by the work of many social documentarians who used the camera as a tool with which to realize their vision and generate change. These photographers actively constructed visual narratives that shaped public opinion and *created* histories, not just recorded them (Abbott, 2010; Goggans, 2010; Harriman & Lucaites, 2007).

Social documentary photographer Dorothea Lange and her husband, economist Paul Taylor, shaped Depression-era narratives by passionately communicating the resilience and pride of American workers in the face of social policies that left them victimized. Lange's iconic photograph *Migrant Mother*, from 1936, recalls High Renaissance portraits of mother and child, but the context belies abject poverty. This dissonance generates a meaningful tension that could only be consciously constructed by the photographer. Her work is a strong example of both moral creativity (Haste, 1993) and the thorny relationship between creative choices and the metaphors of mechanical objectivity, photographic truth, and the passive photographer. Today, Lange is written about as frequently for the content of her work as for the presumed ethical transgressions she committed in making it (Coles, 1997; Goggans, 2010; Harriman & Lucaites, 2007). Lange herself claimed on several occasions that her photographic process was free of any arranging, manipulating, or otherwise tampering with the scene or resulting image (Lange, 1982; Lange, Coles, & Heyman, 2005). However, investigations of her archive have proven that she dramatically altered narrative meaning through her creative decisions before and after image capture. Lange created her work in a culture trained to be more persuaded by the illusion of an objective truth, so admitting to alterations that supported her conceptual ideas and creative objectives would have reduced the credibility of her work. The appropriateness of creativity in documentary work was questioned overall (Hariman & Lucaites, 2007).

The surprising element of the ethical debates around Lange's staging and manipulations is that very little is written about the impact of her images on the photographic subjects themselves. The use of human subjects by documentary photographers is justified in general through simple moral math—the potential to do good outweighs the potential for harm. Lange (1960) claimed that the subject of *Migrant Mother*, whose name she did not know, knew that Lange's photographs would help her situation. When the subject of *Migrant Mother* was identified and interviewed 40 years later, she expressed bitterness that her portrait benefited Lange but not herself (Hariman & Lucaites, 2007). The complicated moral implications of using human subjects are further exacerbated by the assumption that what we see is an objective representation—that the woman we experience in the photograph is the woman who was photographed. Photographers, out of necessity, use individuals to *construct* narrative meaning. Encouraging moral and ethical practices is essential, but as well we need to find metaphors that adequately convey the complexities of photography and the agency of photographer, perceiver, and subject.

The metaphor of the dialogic self

In other domains, the concept of the dialogic self, an epistemological metaphor that describes the transactional nature of individual social engagement, has been used to address embedded Modernist conventions, positivist epistemologies, and monologic thinking. The dialogic self joins sociocultural constructions of the self (for example, Vygotsky, 1978, 1986) and builds from the concept of polyphony (Bakhtin, 1981, 1986), the co-constructive presence of many voices, multiple perspectives, and meanings in all inter- and intrapersonal communicative acts (Haste, 2013). For Bakhtin, creative acts exist in the dialogic relationship between what is ready-made, and what is created from the ready-made by an author/artist and understood by a reader/perceiver: "It is as if everything given is created anew in what is created, transformed in it" (1986, p. 120). Perpetual and co-constructive dialogic engagement is transacted primarily through language with oneself and others, but also with material objects (Bakhtin, 1986; Holquist, 2002; Pechey, 2007). Individual voices are saturated with the voices of others, and thinking is teeming with multiple "I positions" (Hermans, 1996) engaged from different viewpoints.

Important to overcoming metaphors of photographic objectivity and truth, Bakhtin's rejection of a monologic perspective does not reject the possibility of truth, but claims that "a single essential truth is possible if truth is understood in a fundamentally dialogical way" (Salgado & Clegg, 2011, p. 433). Bakhtin writes: "It is completely possible to imagine and to assume that this one and only truth requires a plurality of consciousnesses, and that it has, so to speak, the nature of an *event* and is born in the point of contact of various consciousnesses" (Bakhtin, 1973, p. 65, emphasis in original; also Salgado & Clegg, 2011, p. 433). Building truth from a multiplicity of truths is an appropriate and promising perspective in the context of globally networked documentary image-making. One hope for the domain is that the citizenry of photography (Azoulay, 2012) contributes to the polyphony of perspectives that engage viewers in a dialogue of active selecting, comparing, and interpreting (Ranciere, 2009).

Fundamental to dialogic self theory current in sociocultural and cultural psychology is the valuation of the unique contribution of the individual but contestation of individual autonomy. Theorists focus on the dynamic through which individuals are shaped by and act with meditational means provided by the culture (for example, Haste, 1993, 2008, 2013; Richardson, 2011; Rogoff, 2008; Vygotsky, 1978, 1986; Wertsch, 1985, 1991). The individual is not merely influenced by cultural tools but is an active agent in dialogue with those resources, both cultural and interpersonal, drawing on them but also discriminating, interpreting, altering, and contributing. We actually constitute our world through negotiation, interpretation, and engagement (Haste, 2013). Through this lens, the camera is a powerful mediating tool that both constrains and facilitates what the photographer is able to create. The photographer is neither dominated by nor dominant over the camera, but relationally engaged. This perspective is a significant counter-argument to the truthful mechanical eye of the camera.

The cultural approach

Approaches to diverse areas of inquiry have been influenced by the combined work of Lev Vygotsky (1978, 1986), James Wertsch (1985, 1991), and Mikhail Bakhtin (1981, 1986). Scholars of artistic creativity (Glaveanu, 2010, 2011; Richardson, 2011), moral development (Haste, 1993, 1994, 2008; Tappan, 2005, 2006), and civic engagement (Haste, 2009) have engaged this foundation to create sociocultural/dialogic models of their disciplines. These approaches, built from the same foundation, dovetail easily and present a "distributed, collective, shared, fundamentally *dialogical* view ... that stands in contrast to the individualistic, atomistic, isolated, fundamentally *psychological* view that has dominated the field" (Tappan, 2006, p. 15, emphasis in original). These sociocultural/dialogic models provide examples from which to build a theory specific to photography that is in alignment with the complicated working context of documentary photographers.

Vlad Glaveanu's (2010, 2011, 2013) work on creativity uses a Vygotsky, Bakhtin, and Wertsch theoretical perspective to shift the focus of investigation away from the creative individual and toward the exploration of different *patterns* of engagement between the individual, others, and cultural resources. Glaveanu (2010, 2011) proposes a tetradic framework in which creativity occurs through the interconnection of self (creator), others (community), new "artifact" (creative product/process), and existing artifacts (previous knowledge and practices). Glaveanu's dimension of the new artifact identifies the ongoing process of making as a component engaged in dialogue with self–others and existing artifacts, but does not engage the materials at work in the process as a separate dialogic dimension. The metaphor of dialogue with materials appears frequently in visual artists' descriptions of the creative process (for example, Kesten, 1997; Shahn, 1957; Weschler, 2008a, b) and the findings of those who study them (for example, Bamberger & Schon, 1983; Eisner, 1983, 2002; Gayford, 2010; Kimmelman, 1998). Models for better describing and understanding the working process of photographers need to consider materials that "talk back" in dialogue with their co-creator.

The importance of the material component of creativity is identified in a recent study introducing a theory of the sociomateriality of creativity (Tanggaard, 2012). This theory highlights the "close relationship between human beings and material tools in the creativity process" (2013, p. 21) and the degree to which connecting to or resisting those materials shapes creative possibilities, decisions, and outcomes. Although Tanggaard (2012) does not build his argument on a tool–user foundation, this thinking complements both the tool–user model (Wertsch, 1985, 1998; also Haste, 2008, 2009; Tappan, 2005, 2006) and Bakhtin's (1986) individual–material dialogues.

Photographers working in documentary traditions that build meaning from the focal subject of the image describe the interconnection of their creative process and the subject(s) of a particular situation. Accounts of this mode of creative practice need to take into account not just the interconnection between the self, social, existing artifacts, and the new artifact (Glaveanu, 2010, 2011), but also the photographer's *material*—the focal subject of the lens.

The dialogic witness

Building on cultural theories of the dialogic self, the cultural triangle (Haste, 2008, 2009, 2013), and existing models of creativity and moral development, it is possible to construct a model of documentary photographic practice. Drawing on the language of photography but subverting it with reference to the dialogic self, the dialogic witness emerges as a new metaphor for creativity in the domain. In this model, the creative process is transacted through the dynamic interaction of four dimensions: individual, symbolic, social, and situational (Figure 13.1).

Underpinning this model is the cultural triangle (Haste, 2008, 2009, 2013), which describes individual functioning as interconnected, co-constructive dialogue between intrapersonal sense-making (the individual), interpersonal dialogue (others), and use of mediational tools drawn from the culture. Cultural psychology regards the individual as a tool-user, a model of analysis that frames all action as mediated through dialogic engagement with cultural tools and resources (for example, Haste, 2008, 2009; Vygotsky, 1978; Wertsch, 1991). Individuals are both shaped by and shaping what is possible with those tools. This relationship highlights the connection between the individual and

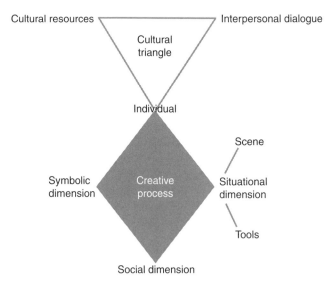

Figure 13.1   Model of the dialogic witness.

the tool, and stresses the existence of co-constructive dialogue between tool and user (Haste, 2008, 2009; Wertsch, 1998). Individual–tool relationships vary by individual, by tool, and by context, but cultural psychologists view the interaction of individual and tool to be engaged and dynamic, not passive. A cultural analysis of the overarching narrative of photography suggests an underlying perception that subverts the individual–tool relationship. In many ways, photographic metaphors suggest that the tool uses the individual, not the other way around, imposing a false and unnecessarily limiting constraint on the medium.

Building from the individual dimension of the cultural triangle, the model of the dialogic witness posits that the creative process of the individual photographer engages in a co-constructive dynamic that is transacted dialogically between the following four dimensions.

Symbolic dimension

The symbolic dimension includes the individual's archive of visual references, conceptual and practical understandings of the medium, and knowledge of the domain (see Moran, 2009). In the domain of photography, cultural tools of the medium consist of various stylistic genres that create endless reservoirs of visual resources, references of quality, examples of aesthetic value, and exemplars within and across genres (like Riis, Hine, or Lange).

Social dimension

The social dimension includes the professional field, the ethical and creative demands of the media outlet or other destination for the individual's work, and perceived expectations of the intended audience. While moral considerations are governed on the individual level and ethical considerations governed on the social level (Moran, 2009), both ethical and moral issues are informed by individual and social dialogue.

Situational dimension

The situational dimension consists of two interactive forms of material: tools and the referent or object of focus. Photographic tools are the many forms of equipment and devices for image capture, processing, and production. The photographer is in dialogue with these tools, constrained by them, but also manipulating them to service the individual's emergent creative dialogue with the scene. This dynamic has been characterized as improvisation by practitioners (Klein, 2010; Light, 2000), but these represent lay theories of improvisation, not theories current in the study of creativity (for example, Fisher & Amabile, 2009). The increased interconnection of photography and communication makes the relationship between the individual and the situational dimension of the creative process of particular interest. It is through this dynamic dimension that the individual engages with the physical world and the living subjects in front of the lens while in the process of creating. That individual's creative response is contingent on receptivity to and efficacy with those situational dialogues. Dorothea Lange (Lange, Coles, & Heyman, 2005) wrote of the importance of the receptive eye, an eye that is able to receive without prejudice the visual information as it unfolds. A cultural account of this receptivity would deny that any interaction can be unprejudiced by preexisting knowledge and assumptions, but Lange's point suggests she was approaching her subjects with a give-and-take mirroring dialogue.

The benefits of fresh metaphors

This chapter suggests a model that would enable researchers to study documentary photographers as active participants constructed by and constructing cultural narratives through dialogic engagement with the tools of their medium. This process involves the interplay of visual perception, contextual understanding, affective response, narrative interpretation, aesthetic preferences, moral imperatives, awareness of audience, and expectations for exhibition platform. Applying this model to the work of practicing photographers will undoubtedly enrich our appreciation of how their complex creative process is both shaped by and shapes the world around us. As more research surfaces using cultural approaches in the work of photographers, the field will be liberated from the false expectation of an indexical relationship to an external reality, and photographic work will be evaluated on the merits of the photographer's insights, not its authenticity (Aker, 2012; Ritchin, 2009). Disengaging metaphors of truth and reality from photographic making and perceiving encourages better ethical and creative practice, but also allows viewers to co-create narrative meaning.

One of the difficulties of writing about photography is that "[t]here is no single or simple story to tell" (Heifferman, 2012, p. 11). There are so many forms of image-making made for countless purposes and according to principles of a variety of discourses, from media to art to science to the law. The porous nature of boundaries within the domain allows images to function simultaneously in multiple discourses. An image produced in the context of photojournalism can pass through the channels of documentary picture story and into the exhibition platform of art.

Documentarians can aestheticize their style, and artists can use a documentary style. Amateurs publish professionally, and professionals make personal snapshots. It is a complex domain that has produced a turbulent field. One issue that runs through the various discourses is the relationship of photography to what is "real" or "true" (Thompson, 2003). Artists working in the medium often play on its relationship to truth to subvert the viewer's expectation of reality (for example, Jeff Wall, Loretta Lux, Julie Blackmon) while, as explained in this chapter, many journalists and documentarians still rely on it to anchor the "objectivity" of their images.

Explicating metaphors embedded in the domain, identifying their impact on the field, and suggesting a deeper understanding of creative and ethical work in photography are important to the health of the domain. Photography is a powerful tool that can and should be used in acts of moral creativity (Haste, 1993). Globally networked photographic devices have tremendous creative potential to influence issues of moral concern. The complex civic and social consequences of a networked global community are already apparent: the mass mobilization seen during the Arab Spring; the development of agency in disempowered groups linked through social networks; and grassroots issues gaining national attention through blogging (Haste, 2008, 2009). As the next generation comes of age with these technologies, it is important to develop metaphors and understandings, like the dialogic witness suggested in this chapter, that better represent the complex psychological and social demands of balancing creative agency and ethical action.

References

Abbott, B. (2010). *Engaged Observers: Documentary Photography since the Sixties.* Los Angeles, CA: J. Paul Getty Museum.

Aker, P. (2012). Photography, objectivity and the modern newspaper. *Journalism Studies 13*(3), 325–339.

Azoulay, A. (2012). *Civil Imagination: A Political Ontology of Photography.* New York: Verso.

Bakhtin, M. M. (1973). *Problems of Dostoevsky's Poetics* (R.W. Rotsel, trans.). Ann Arbor, MI: Ardis.

Bakhtin, M. M. (1981). *The Dialogic Imagination.* Austin: University of Texas Press.

Bakhtin, M. M. (1986). *Speech Genres and Other Late Essays.* Austin: University of Texas Press.

Bamberger, J., & Schon, D. (1983). Learning as a reflective conversation with materials: Notes from work in progress. *Art Education, 36*(2), 68–73.

Barthes, R. (1981). *Camera Lucida.* New York: Hill and Wang.

Batchen, G., Gidley, M., Miller, N., & Prosser, J. (2012). *Picturing Atrocity: Photography in Crisis.* London: Reaktion Books.

Billig, M. (1995) *Arguing and Thinking*, 2nd edn. Cambridge: Cambridge University Press.

Bogre, M. (2012). *Photography as Activism*. Waltham, MA: Focal Press.

Chapnick, H. (1994). *Truth Needs No Ally*. Columbia: University of Missouri Press.

Coles, R. (1997). *Doing Documentary Work*. New York: Oxford University Press.

Eisner, E. W. (1983). On the relationship of conception to representation. *Art Education, 36*(2), 22–27.

Eisner, E. W. (2002). *The Arts and the Creation of Mind*. New Haven, CT: Yale University Press.

Fisher, C., & Amabile, T. (2009). Creativity, improvisation and organizations. In T. Rickards, M. Runco, & S. Moger (eds.), *The Routledge Companion to Creativity* (pp. 13–24). London: Routledge.

Flusser, V. (1983). *Towards a Philosophy of Photography*. Glasgow: Reaktion Books.

Foster, M. D. (2009). What time is this picture? Cameraphones, tourism, and the digital gaze in Japan. *Social Identities: Journal for the Study of Race, Nation and Culture, 15*(3), 351372.

Freund, G. (1982). *Photography and Society*. Boston, MA: David Godine.

Gardner, H. (2006). Creativity, wisdom, and trusteeship. In A. Craft, H. Gardner, & G. Claxton (eds.), *Creativity, Wisdom, and Trusteeship: Exploring the Role of Education* (pp. 49–65). Thousand Oaks, CA: Corwin Press.

Gayford, M. (2010). *Man with a Blue Scarf: On Sitting for a Portrait by Lucien Freud*. London: Thames & Hudson.

Glaveanu, V. (2010). Principles for a cultural psychology of creativity. *Culture and Psychology, 16*(2), 147–163.

Glaveanu, V. (2011). How are we creative together? Comparing sociocognitive and sociocultural answers of knowledge. *Theory and Psychology, 21*(4), 473–492.

Glaveanu, V. (2013). Rewriting the language of creativity: The five A's framework. *Review of General Psychology, 17*(1), 69–81.

Goggans, J. (2010). *California on the Breadlines: Dorothea Lange, Paul Taylor and the Making of a New Deal Narrative*. Berkeley: University of California Press.

Hariman, R., & Lucaites, L. (2007). *No Caption Needed: Iconic Photographs, Public Culture, and Liberal Democracy*. Chicago, IL: University of Chicago Press.

Haste, H. (1993). Moral creativity and education for citizenship. *Creativity Research Journal, 6*(1–2), 153–164.

Haste, H. (1994). *The Sexual Metaphor*. Cambridge, MA: Harvard University Press.

Haste, H. (2008). Constructing competence: Discourse, identity and culture. In I. Plath (ed.), *Kultur – Handlung – Demokratie: Diskurse ihrer Kontextbedingungen* (pp. 109–134). Wiesbaden: Verlag fur Sozialwissenschaften.

Haste, H. (2009). What is "competence" and how should education incorporate new technology's tools to generate "competent civic agents"? *Curriculum Journal, 20*(3), 207–223.

Haste, H. (2011). Discovering commitment and dialogue with culture. *Journal of Moral Education, 40*(3), 369–376.

Haste, H. (2013). Culture, tools and subjectivity: The (re)construction of self. In T. Magioglou (ed.), *Culture and Political Psychology: A Societal Perspective*. Charlotte, NC: Information Age Publishers.

Heifferman, M. (2012). *Photography Changes Everything*. New York: Aperture Foundation.

Hermans, H. J. M. (1996). Opposites in a dialogical self: Constructs as characters. *Journal of Constructivist Psychology, 9*(1), 1–26.

Holquist, M. (2002). *Dialogism,* 2nd edn. London: Routledge.

Keats, P. (2010). The moment is frozen in time: Photojournalists' metaphors in describing trauma photography. *Journal of Constructivist Psychology, 23*(3), 231–255.

Kesten, J. (1997). *The Portraits Speak: Chuck Close in Conversation with 27 of His Subjects.* New York: ART Press.

Kimmelman, M. (1998). *Portraits: Talking with Artists at the Met, the Modern, the Louvre and Elsewhere.* New York: Random House.

Klein, A. (2010). *Words without Pictures.* New York: Aperture Foundation.

Lakoff, G., & Johnson, M. (2003). *Metaphors We Live By.* Chicago, IL: University of Chicago Press.

Lange, D. (1960). The assignment I'll never forget: Migrant Mother. *Popular Photography, 46*(Feb.), 42–43.

Lange, D. (1982). *Photographs of a Lifetime: A Monograph.* New York: Oxford University Press.

Lange, D., Coles, R., & Heyman, T. (2005). *Dorothea Lange: Photographs of a Lifetime.* New York: Aperture Foundation.

Light, K. (2000). *Witness in Our Time.* Washington, DC: Smithsonian Press.

Linfield, S. (2010). *The Cruel Radiance: Photography and Political Violence.* Chicago, IL: University of Chicago Press.

Moller, F. (2012). Celebration and concern: Digitization, camera phones and the citizen-photographer. In C. Martin & T. von Pape (eds.), *Images in Mobile Communication* (pp. 57–78). Wiesbaden: VS Research.

Moran, S. (2009). Creativity: A systems perspective. In T. Rickards, M. Runco, & S. Moger (eds.), *The Routledge Companion to Creativity* (pp. 292–301). London: Routledge.

Morris, E. (2011). *Believing Is Seeing: Observations on the Mysteries of Photography.* London: Penguin Press.

Newton, J. H. (2001). *The Burden of Visual Truth: The Role of Photojournalism in Mediating Reality.* Mahwah, NJ: Lawrence Erlbaum Associates.

Pechey, G. (2007). *Mikhail Bakhtin: The Word in the World.* London: Routledge.

Ranciere, J. (2009). *The Emancipated Spectator.* London: Verso.

Reading, A. (2009). Mobile witnessing: Ethics and the camera phone in the "War on Terror." *Globalizations, 6*(1), 61–76.

Richardson, F. C. (2011). A hermeneutic perspective on dialogical psychology. *Culture and Psychology, 17*(4), 462–472.

Ritchin, F. (2010). *After Photography.* New York: Norton.

Ritchin, F. (2012). Toward a hyperphotography. In G. Batchen, M. Gidley, N. Miller, & J. Prosser (eds.), *Picturing Atrocity: Photography in Crisis.* London: Reaktion Books.

Rogoff, B. (2008). Thinking with the tools and institutions of culture. In P. Murphy & K. Hall (eds.), *Learning and Practice: Agency and Identities* (pp. 49–70). Los Angeles, CA: Sage.

Rosler, M. (1992). In, around and afterthoughts (on documentary photography). In R. Bolton (ed.), *The Contest of Meaning* (pp. 304–334). Boston, MA: MIT Press.

Salgado, J., & Clegg, J. (2011). Dialogism and the psyche: Bakhtin and contemporary psychology. *Culture and Psychology, 17*(4), 421–440.

248   *The Ethics of Creativity*

Schaaf, L. (2000). *The Photographic Art of William Henry Fox Talbot*. Princeton, NJ: Princeton University Press.

Schiller, D. (1977). Realism, photography and journalistic objectivity in 19th century America. *Studies in the Anthropology of Visual Communication, 4*(2), 86–98.

Sekula, A. (1982). Photography between labor and capital. In B. Buchloh & R. Wilkie (eds.), *Mining Photographs and Other Pictures: Photographs by Leslie Shedden* (pp. 193–268). Halifax, Nova Scotia: Press of the Nova Scotia College of Art and Design.

Sekula, A. (2006). The body and the archive. In C. Merewether (ed.), *The Archive* (pp. 70–75). Cambridge, MA: MIT Press.

Shahn, B. (1957). *The Shape of Content*. Cambridge, MA: Harvard University Press.

Soffer, O. (2009). The competing ideals of objectivity and dialogue in American journalism. *Journalism, 10*(4), 473–491.

Sontag, S. (1966). *Against Interpretation and Other Essays*. New York: Farrar, Strauss & Giroux.

Sontag, S. (1990). *On Photography*. New York: Anchor Books/Doubleday.

Sontag, S. (2003). *Regarding the Pain of Others*. New York: Picador.

Stern, M. (2012). Presence, absence, and the presently-absent: Ethics and the pedagogical possibilities of photographs, educational studies. *Journal of the American Educational Studies Association, 48*(2), 174–198.

Tanggaard, L. (2012). The sociomateriality of creativity in everyday life. *Culture and Psychology, 19*(1), 20–32.

Tappan, M. (2005). Mediated moralities: Sociocultural approaches to moral development. In M. Killen & J. Smetana (eds.), *Handbook of Moral Development* (pp. 351–374). Hillsdale, NJ: Lawrence Erlbaum Associates.

Tappan, M. (2006). Moral functioning as mediated action. *Journal of Moral Education, 35*(1), 1–18.

Thompson, J. (2003). *Truth and Photography*. Chicago, IL: Ivan R. Dee.

Trachtenberg, A. (1990). *Reading American Photographs: Images as History, Mathew Brady to Walker Evans*. New York: Hill and Wang.

Tremain, K. (2000). Introduction: Seeing and believing. In K. Light (ed.), *Witness in Our Time* (pp. 1–11). Washington, DC: Smithsonian Press.

Vygotsky, L. S. (1978). *Mind in Society: The Development of Higher Psychological Processes*. Cambridge, MA: Harvard University Press.

Vygotsky, L. S. (1986). *Thought and Language* (Alex Kozulin, ed.). Cambridge, MA: MIT Press.

Wertsch, J. (1985) *Vygotsky and the Social Formation of Mind*. Cambridge, MA: Harvard University Press.

Wertsch, J. (1991). *Voices of the Mind: A Sociocultural Approach to Mediated Action*. Cambridge, MA: Harvard University Press.

Wertsch, J. (1998). *Mind as Action*. New York: Oxford University Press.

Weschler, L. (2008a). *Seeing Is Forgetting the Name of the Thing One Sees: Over Thirty Years of Conversations with Robert Irwin*. Berkeley: University of California Press.

Weschler, L. (2008b). *True to Life: Twenty-Five Years of Conversations with David Hockney*. Berkeley: University of California Press.

West, N. M. (2000). *Kodak and the Lens of Nostalgia*. Charlottesville: University Press of Virginia.

Wolf, C., & Hohfeld, R. (2012). Revolution in journalism? In C. Martin & T. von Pape (eds.), *Images in Mobile Communication* (pp. 81–99). Wiesbaden: VS Research.

Zelizer, B. (2010). *About to Die: How News Images Move the Public*. Oxford: Oxford University Press.

14
Neglect of Creativity in Education: A Moral Issue

Arthur Cropley
University of Hamburg, Germany

The problem

Teaching is not just a job like any other. Teachers belong to the only occupational group whose core responsibility is mentoring the next generation and shaping the future of the societies of which they are part. Thus, teachers have a special duty to their charges and to society. By entering the profession teachers accept the responsibilities associated with this duty, whether they consciously seek them or not. They thus have *moral* obligations that are different in principle from the manual, technical, or professional obligations of most other occupations. This chapter will examine the dimensions of these moral obligations. Its purpose is not to cast teachers in the role of the villain or to castigate them for their moral failings, but to examine promotion/lack of promotion of creativity in the classroom from a rarely discussed perspective in order to gain new insights into the situation.

It is now widely accepted that creativity is crucial for the economic, social, and personal welfare of society. This issue will be addressed in more detail below. Pink (2005) summarized what is needed in striking terms: technologically advanced societies have now reached the stage where the mere ability to store knowledge and make logical connections and draw obvious, correct conclusions can easily be done by information technology. What is needed now is for society to advance beyond the "Information Age" into the "Conceptual Age." In the cognitive domain this would involve *synthesizing*: seeing relationships among apparently unrelated pieces of information, making unexpected combinations of ideas, detecting broad patterns, and similar kinds of thinking. Personality dimensions and motivation that would support such thinking include openness, resistance to social pressure, tolerance

250

for ambiguity, risk-taking, or willingness to delay closure, to give some examples.

These are processes and personality characteristics typically associated with creativity in psychological discussions (for example, Cropley & Cropley, 2009). Unfortunately, there is evidence, which is discussed in greater detail in a later section, that the very processes and personal properties called for by Pink are becoming *less* well developed in modern classrooms. This tendency is not confined to the USA, but has been widely reported in other countries too (for a summary of some Australian discussions, see Cropley, 2012). This would be a matter of no great concern if creativity were about as important as, let us say, snorkeling or speaking Latin: both are good fun and extremely important to subgroups whose members rightly wish to see the activity promoted and fostered, but they are not crucial for the welfare of the entire society. However, creativity is, and because of the special and particular responsibility of teachers to promote the wellbeing of society, any failure on their part to promote creativity is a moral issue. Of course, there are many honorable exceptions. Furthermore, as already stated the purpose of this chapter is not to *attack* teachers, vilify them, or label them unethical, but to demonstrate that the whole problem field of the promotion of creativity in schools is not simply an organizational, curricular, pedagogical, or policy matter, but, because of the combination of conditions presented above (teachers have a special responsibility to promote the wellbeing of the society and creativity is a vital element in this wellbeing), the field of education has striking moral responsibilities and needs to be analyzed from this perspective. (Elsewhere—for example, Cropley, 2012—I have summarized the role of creativity as a pedagogical principle.)

What is morality?

As the term is used in this chapter, morality involves ideals about what is right and what is wrong, often derived from the teachings of great thinkers. Rightness/wrongness at an everyday level, such as crossing an intersection on a red light, is not usually regarded as a moral issue. Rather, the term is reserved for rightness/wrongness at a higher level such as goodness/badness, justice/injustice, or honesty/perfidy. Peterson and Seligman (2004) showed that certain values are regarded as right in many societies and have been for millennia, so that there may be universal rights and wrongs. According to Plato, for instance, there are four "cardinal" virtues (courage, temperance, prudence, and justice) that should guide all actions. All other virtues are thought to spring from these. However, in this chapter moral rightness and moral wrongness are not regarded as absolutes, but as deriving from a specific system of beliefs, especially cultural norms; morals are thus seen here as descriptive rather than prescriptive (that is, they refer to a particular society and, indeed, may change within a given society with the passage of time). What is moral in one society at a particular time (such as polygamy in some societies) may be immoral according to the norms of others (such as those that are monogamous), or even at a different time in the same society. Thus, morality has a strong social dimension.

A sister topic to morals is ethics. Ethics are concerned with how a moral person should *act* (as against what a moral person should believe is right). Ethical behavior involves *doing* what is morally right and proper, not simply what is expedient. Kant argued that there is a moral imperative to treat other people with dignity, and never to exploit them as instruments of one's own satisfaction. According to John Locke, no one should act in such a way as to harm anyone else's life, health, liberty, or possessions. Bekoff and Pierce gave an idea of the social content of moral behavior. According to them, morality involves "a suite of inter-related other-regarding behaviors that cultivate and regulate complex interactions *within social groups*" (2009, p. 7, emphasis added). Morality is concerned, according to them, with phenomena such as altruism, trust, compassion, or fair play. This definition directs attention squarely to the social nature of morals and, especially, to the central role of concern for the wellbeing of others (morals are "other-regarding"). Thus, moral behavior involves a responsibility to act in ways that avoid doing harm to others and *to promote the common good*.

The idea of the common good may seem problematic here. Technologically advanced wealthy countries are becoming more and more diverse culturally as a result of the massive inflow of people from countries where traditions and norms differ from those of longer-established majorities in the receiving societies. However, "common good" is understood here in a much broader way than as narrow, specific rules on concrete issues such as whether having many wives is a good thing or not. Indeed, any attempt to define the common good in terms of unchanging concrete principles would run directly counter to the core idea of creativity: change and renewal. Henning (2005) worked out a number of obligations for moral behavior from the writings of Alfred North Whitehead, and these help to understand what is meant here by the "common good." The obligations include bringing about the widest possible universe of beauty, maximizing the intensity and harmony of oneself and everything within one's sphere of influence, and avoiding destruction. Matters such as altruism, trust, compassion, or fair play, which have already been mentioned, also transcend geographically and temporally changing ideas of right and wrong, unless beauty, harmony, avoidance of harm, altruism, compassion, and fair play are regarded as passing fads.

The social nature of morality and ethical behavior means that, by virtue of their role in society, teachers have a particularly strong moral responsibility, and any failure to promote creativity has a clear moral/ethical dimension, whereas in other groups, such as auto repairers or even lawyers, failure to do what the practitioners know is right might be no more than a sign of less than optimal technical or professional competence. Indeed, in some artistic circles acting in ways that are regarded as socially reprehensible may even be regarded as admirable (see the example given below). There is now widespread acceptance that creativity is good for both society as a whole and also individuals. In the case of teachers, research has shown that as a group they know this, and even express their support for creativity when asked what they think of it. Despite this, research in many countries and over many years has repeatedly shown that in actual classroom practice a majority of teachers dislike creative children and make little effort to foster creativity, despite their familiarity with and acceptance of its value. For the reasons outlined above, this paradoxical situation (teachers know what is good, but fail to do it) is a moral issue. The acquiescence of policy-makers and teacher educators in teachers' failure to implement creativity-based teaching also has a moral aspect. Of course, most people, if not all, sometimes fail do what they know they ought to, but in the case of teachers and creativity the failure to do what is right and good takes on special moral contours.

When people knowingly act against the common good or even when they knowingly fail to act for the common good, their behavior is immoral. One classic immoral situation occurs when a person (or group of people) knows that some course of action is wrong (according to the ethical system of the society in which they live or the code they follow), but persists in the course of action anyway, for instance for personal gain (such as the acquisition of wealth or power). This situation is also seen in a less obvious form when the morally wrong behavior is pursued in order to fit in with a subgroup, or even because it is simply easier to follow the morally wrong course—it leads to less "hassle" and yields a quiet life. People who turn a blind eye to the persecution of minorities because everyone is doing it, despite the fact that they know that this is wrong, would be an example. Thus, indifference to moral issues is also in itself a moral issue.

In a direct discussion of creativity and morality, Martin made the link to social factors explicit by pointing out that morality should be looked at *in context*. He focused on four social contexts (2006, p. 55): role responsibilities, leadership, policy advising, and authenticity. It is immoral for people with responsibilities arising from their role in society, people who are leaders in the community, and people involved in making policy to fail to advance the general good. The first three of these contexts are highly relevant to the work of teachers—they bear heavy responsibility associated with their role as nurturers of the next generation, are expected to display both personal and academic leadership in the classroom, and are strongly involved in providing feedback on policy. Thus, teachers operate in a context in which moral issues are of special importance.

The benefits of creativity

In order to investigate the moral dimensions of the way many teachers behave vis-à-vis creativity, it is necessary to show that (1) creativity is a strong force for the common good; (2) teachers know this or ought to know it (after all, they have received intense professional training from experts); and (3) they are knowingly not fostering it. Creativity is understood here in the now familiar way referred to by Runco and Jaeger (2012, p. 92) as the "standard definition." It involves the generation of *useful novelty*. According to the "Four Ps" model of creativity summarized by Cropley and Cropley (2009), the generation of relevant and effective novel *products* depends on appropriate thinking (creative *process*) and personal properties (creative *person*), and is facilitated/inhibited by a particular environment (*press*). According to Csikszentmihalyi (1996), humankind's fate now depends on creativity. Prompted by new and unprecedented difficulties and dangers at a global level, such as the global financial crisis or global warming, many writers now see creativity as the force that will rescue the economy and save the natural environment (for example, Martin & Florida, 2009; Pink, 2005). In the social domain, Oral argued that creativity is vital "for shaping... future orientations and actualizing reforms in political, economic and cultural areas" (2006, p. 65). At the level of the individual, Plucker, Beghetto, and Dow (2004) documented research findings on the positive contribution of creativity to individual wellbeing both in the intrapersonal domain (for example, good mental health, effective coping, emotional growth, success in therapy) and the interpersonal domain (for example, workplace leadership, life success, maintenance of loving relationships, conflict resolution, and more satisfying interpersonal relationships).

Creativity is also known to have positive effects in the classroom itself; that is, not as a *goal* or hoped-for *outcome* of teaching but as a pedagogical tool for improving the *process* of pupils' learning in a wide variety of subject areas. Cropley (2012) summarized research showing that creative pedagogy leads not only to better grades, but also to benefits that are not directly part of academic work at all (for example, reduced likelihood of dropping out of school early, less boredom, and a more positive self-concept), learning to work cooperatively, improved planning and goal setting, developing the ability to make and accept constructive criticism, better expression of emotions, and possessing better life skills (for example, conflict resolution, stress management, and empathy). Such effects have also been seen with "disempowered and disenfranchised" students. Benefits more directly related to creativity included improvement in cognitive processes such as combining elements, lateral thinking, problem definition, and idea generation, as well as personal properties such as imaginativeness, willingness to take risks, or openness to the new.

Teachers' neglect of creativity

Turning to the second issue, early research by Feldhusen and Treffinger (1975) showed that 96 percent of the teachers they surveyed agreed that creativity is a good thing, and Runco, Johnson, and Bear (1993) reported similar favorable attitudes to creativity among teachers (in principle). However, the fact is that in real life teachers not infrequently disapprove of or even dislike the students in their classes who are most creative (Dawson et al., 1999), and many do little to foster creativity. As Westby and Dawson (1995) showed, many teachers who claimed to have a favorable view of creative children almost bizarrely described them as "conforming." When the teachers in the Westby and Dawson study were given adjectives describing traits typical of what creative children are really like (for example, risk-taking, curious), they said they *disliked* such youngsters; see also Aljughaiman and Mowrer-Reynolds (2005). As Smith and Carlsson (2006, p. 222) put it, "teachers seem to have a confused picture of what is a favorite pupil and what is a creative pupil."

This finding is not new. Early studies (for example, Holland, 1959; Torrance, 1959) concluded that highly creative children tended to be less well known and less well liked by teachers than did children with a high IQ. Getzels and Jackson (1962) reported the results of a study in which teachers expressed a strong preference for teaching "merely" intelligent children (as against creative youngsters), and this was not attributable to superior classroom performance, as there were no differences in achievement between the two groups. The finding is also not confined to the United States. In an early study going beyond the USA to include Germany, India, Greece, and the Philippines, Torrance (1965) showed that there was near unanimous disapproval of creative schoolchildren across countries. Teachers and parents in India reported favorable views of creativity, but also linked with it several words associated with mental illness (emotional, impulsive; Runco & Johnson, 2002). Chan and Chan (1999) found that Chinese teachers associated socially undesirable traits such as arrogance and rebelliousness with creativity in students.

In fact, over the years and right up to today, similar findings have consistently been reported in a number of different countries and regions, including Africa (Obuche, 1986), Australia (Morgan & Forster, 1999), Chinese societies (Chan & Chan, 1999; Tan, 2003), Europe (Brandau et al., 2007; Karwowski, 2007), the Middle East (Oral & Guncer, 1993), and North America (Dawson et al., 1999; Scott, 1999). In reviewing the situation, Makel came to the disturbing conclusion that "we have deemed creative development and performance extremely important in the professional world of adults, but appear to minimize it in children" (2008, p. 38).

The situation is similar in higher education. A survey of employers in Australia (Commonwealth of Australia, 1999) indicated that three-quarters of new graduates in that country were "unsuitable" for employment because of "skill deficiencies" in creativity, problem-solving, and independent and critical thinking. This conclusion was supported by Tilbury, Reid, and Podger, who also reported on an employer survey in Australia that concluded that Australian graduates lack *creativity* (2003, p. viii). In the UK, Cooper, Altman, and Garner (2002) concluded that the system discourages innovation; the British General Medical Council, for instance, recognized that medical education is overloaded with factual material that discourages higher-order cognitive functions such as evaluation, synthesis, and problem-solving, and engenders an attitude of passivity.

Although the European Union has established programs bearing the names of famous innovators such as SOCRATES or LEONARDO, it is astonishing that in the guidelines for the development of education in the community, concepts like *innovation* or *creativity* are not prominent.

At least until recently, the Max Planck Institute for Human Development, Germany's leading research institute for the development of talent in research in the social sciences, had never supported a project on the topic of creativity or innovation. In a letter dated April 26, 2006, the office of the President of the Max Planck Society confirmed that the organization did not see creativity as a significant area of research. In the USA, the 1996 report of the Alliance of Artists' Communities concluded that "American creativity" is "at risk." Cropley and Cropley (2005) reviewed findings on fostering creativity in engineering education in the United States and concluded that there is little support for creative students. It is true that there has been some effort in recent years to encourage creativity in colleges and universities. For instance, in 1990, the National Science Foundation established the Engineering Coalition of Schools for Excellence and Leadership (ECSEL), whose goal is transforming undergraduate engineering education. However, a review of current practice throughout higher education in the USA (Fasko, 2000–2001) pointed out that the available information indicated that deliberate training in creativity is rare. In fact, it seems that creativity continues to be neglected: McWilliam and Dawson summarized the situation with almost shocking clarity: "Evocation of 'more creativity' has been limited to rhetorical flourishes in policy documents and/or relegated to the borderlands of the visual and performing arts" (2008, p. 634).

What, then, would morally admirable behavior on the part of teachers involve? There is a very substantial and sometimes controversial literature on the effectiveness of special training in creativity (for a summary see, for instance, Scott, Leritz, & Mumford, 2004). However, Cropley (2012) argued that what is required is to infuse the curriculum with creativity by means of *creative pedagogy* in all disciplines. He claimed that there is evidence that this improves both *outcomes* (greater creativity on the part of pupils and also better learning in general) and also *process* (for example, better motivation, more positive attitudes, greater risk-taking, more daring thinking). He gave examples of what creative pedagogy could involve in elementary school mathematics teaching and secondary school English, while Cropley and Cropley (2009) offered the example of a second-year university class in engineering involving creative pedagogy.

The paradoxical relationship of creativity and society

It seems, then, that teachers are guilty as charged. However, they are not alone. There is, for instance, widespread awareness of the importance of creativity in education at policy-making level, an awareness that is by no means limited to the United States. To take the example of Australia, a number of policy reports reviewed by Cropley (2012) reveal strong awareness of the need for systematic promotion of creativity in the national education system. Despite this, Kim (2011) showed that in the USA, although there are variations across different aspects of creativity and in different age groups, creativity test scores among schoolchildren have *decreased* since 1990, and the decreases are accelerating. Particularly disturbing, perhaps, is that the largest decrease in the ability to produce numerous ideas (fluency) was seen in kindergarten to grade 3, the very age group in which children have traditionally given free expression to their thinking without being inhibited by concerns about accuracy, political correctness, and the like. In addition to scores on creative personality traits such as expressivity, imaginativeness, or unconventionality, scores for cognitive abilities such as fluency (as just mentioned), the ability to produce varied and unusual ideas (originality), or the ability to extend and develop existing ideas (elaboration) have diminished, and are continuing to do so. Motivational properties associated with creativity, such as resistance to premature closure, curiosity, and open-mindedness, are also decreasing. In view of the importance of creativity to the wellbeing of the individual and the society, this constitutes a moral issue. What is happening?

In answering this question it is important to note that aversion to creativity is not confined to teachers and the education system. For instance, DeFillippi, Grabher, and Jones (2007) discussed the paradoxical role of creativity in business and organizations, where it is simultaneously desired and rejected. The evidence is that organizations, scientific institutions, and decision-makers routinely reject creative ideas even while espousing creativity as an important goal (for example, Staw, 1995; West, 2002). In fact, recent research has shown that there is widespread unease about creativity in the broader society, too, despite the almost fawning emphasis on it as something like a universal panacea. Thus, the moral issue of teachers' unwillingness to promote creativity is not evidence of wickedness among teachers, but is part of *a general aversion to creativity in society as a whole*.

This aversion is not, however, simply a result of intolerance or lack of openness in society. Baucus et al. (2008) drew attention to four inherent aspects of the creative process—breaking rules and diverging from standard operating procedures; challenging authority and defying tradition; creating conflict, competition, and stress; taking risks—that are potentially harmful for individual people and society. Thus, paradoxically, creativity is not an unmitigated good. As shown above, it has the potential to make vital contributions to the common good, to be sure, but at the same time it threatens the common good in, among others, the ways just mentioned (breaking rules, challenging authority, creating conflict, taking risks). This is one of many apparently conflicting characteristics of creativity that led Cropley to label it "a bundle of paradoxes" (1997, p. 8). From the point of view of the present chapter, this means not only that failure to foster creativity raises moral issues, as has already been discussed, but that fostering it does too. Put slightly differently, the entire relationship of creativity to society has moral connotations that are not all positive.

The very essence of creativity is challenging what already exists, the way things are currently done, or the way the world is conceptualized. The crucial process through which creativity *threatens* the common good (as against promoting it) and is therefore immoral is by generating *uncertainty* (Mueller, Melwani, & Goncalo, 2012), a state that makes most people uncomfortable. Psychological research (for example, Heimberg, Turk, & Mennin, 2004) has shown that people can only tolerate a limited degree of uncertainty. Beyond a certain critical point it disrupts cognitive functioning, leads to anxiety and even anxiety disorders ranging from simply worrying a lot to clinical states, and encourages obsessive-compulsive behavior. Especially for people in leadership roles, creativity places the value of laboriously acquired knowledge and skills in question, threatens status and moral authority, and may affect people's self-image by, for instance, making them look and feel foolish, intolerant, or rigid. Thus, it is hardly surprising that, despite lip service to the contrary, many people—quite apart from teachers—dislike highly creative individuals or link creativity with disruptiveness or even pathology. Indeed, the negative effects of unbridled creativity may be as morally reprehensible as those of stifled creativity.

Cropley et al. (2010) examined the morally dubious side of creativity in more detail, referring to its "dark side," and Cropley and Cropley (2013) analyzed the link between creativity and crime. Examples given by these authors include an artist who stole body parts from fresh corpses to use them to enhance the effect of his artworks, and a study of scientists working on the development of weapons of mass destruction who were apparently seduced by the excitement of discovery of the new. The artist's actions clearly violated Kant's moral imperative to treat other people with dignity and never to exploit them as instruments of one's own satisfaction, while the scientists violated all of Locke's principles of ethical behavior (avoidance of harm to other people's life, health, liberty, or possessions). Despite this, the main criticism of the artist from within artistic circles was that he worked with plaster of Paris, which is regarded as an inferior medium. The narrow and winding path along the border between creativity and unethical behavior was discussed in detail by Cropley and Cropley (2013).

The moral price of fostering creativity

Returning specifically to education, as Cropley (2009) pointed out, the dark side of in principle admirable characteristics of creative students such as imaginativeness, curiosity, courage, independence, or determination makes it difficult for teachers to distinguish in the classroom between creativity and disorderliness or disruptiveness, or even sheer willful naughtiness. Creative children's apparently strange questions or unexpected answers to teachers' questions, their selection of unusual topics in classroom exercises, homework, or projects, their lower levels of concern about social norms, or their tendency to be self-willed sometimes make them seem to teachers to be "weird," defiant, aggressive, self-centered, or antisocial. If not stopped by the teacher, such behavior may seem to other students to involve tolerating misbehavior, thus encouraging those others to misbehave too. This leads to a belief that creative children are willfully disruptive. This situation is exacerbated by the fact that many teachers see creativity as interfering with the pursuit of academic goals and thus being something for more artistic disciplines; that is, *somebody else's job*, as Aljughaiman and Mowrer-Reynolds (2005) put it.

Thus, although teachers have a moral duty to foster creativity in the interests of the common good, they simultaneously have a moral duty to avoid the harm to the common good (especially in the classroom) that creativity can bring. In fact, there is a price to be paid for fostering creativity in the classroom. Stated in the general way outlined above, the price is uncertainty. Creativity requires as its first property novelty or, as Bruner (1962) put it, "surprise," and surprise of necessity means not knowing what is coming next. The teacher, however, is professionally *required* to know what is coming next, in fact to specify what is coming next. An example of what can happen taken from my own experience in a counseling center in Germany is to be seen in an anecdote related by a teacher of a grade five class. She discovered that a child in her class with whom she had a strained relationship had made up a picture book entitled "The greatest catastrophes of the twentieth century" as a hobby project. Hoping to establish a degree of rapport with him, she asked him to bring his book to school and present his work to the class. He agreed and—standing in front of the entire class—he opened the book. The first picture was of the teacher! The incident involved a mixture of impertinence and wit. The boy certainly achieved surprise and was highly effective in expressing his view of the teacher in a way that was reprehensible and yet almost admirable. His behavior displayed divergent thinking and boundary-breaking, risk-taking and courage, lack of concern about social niceties and self-centeredness, all related to creativity and, in principle, having high potential for good. However, it also involved personal humiliation for the teacher and questioned her competence (that is, it offended against the moral principles of compassion or avoiding harming the dignity of another person, and in this sense was morally reprehensible). To the other children it seemed like an open challenge to the teacher's authority. She was taken completely by surprise and was filled with uncertainty about how to react, a state that she, not surprisingly, experienced as unpleasant. Her initial reaction was to take the boy to the principal and demand disciplinary action, despite the fact that it was she who had asked him to show his book, and had thus brought the whole thing on herself (that is, her uncertainty and humiliation made her reaction rigid, authoritarian, and stereotyped, whereas flexibility and insight were what was called for). Later, she felt guilty about this—the client in the counseling center was not the boy but the teacher.

In fact, students' creativity can arouse uncertainty in teachers in the *cognitive* domain, by placing what they know, their stock in trade, in question, and thus casting doubt on their knowledge and undisputed intellectual authority. In doing this, it also challenges their *self-image*; for instance, as a highly competent professional who is trained to deal in the most constructive way with events in the classroom and can draw on rich experience to achieve this. It also has *social* effects. For example, the other children in the class may regard the teacher as tolerating indiscipline or playing favorites with a small group of "odd" or "different" children. Parents, too, add to the uncertainty. They often adhere to a traditional view of school learning and good classroom discipline, and may believe that creativity belongs exclusively in lessons in art, drama, or music, thus actually opposing what has been presented here as a moral obligation of teachers, although not of course with unethical intentions. All of this means that pursuing the morally required goal of fostering creativity for its ultimate long-term benefit to the common good simultaneously involves tolerating or even encouraging behavior that can be harmful to the common good in the short term in the ways just indicated. This is the price that must be paid for doing what is morally right.

Dealing with the moral paradox of creativity in the classroom is in large part a social, policy, and pedagogical matter. Providing guidelines for what is needed in these areas would go far beyond the limits of this chapter. However, looking at the issues from the point of view of their moral dimensions offers an approach to the whole area that is capable of opening up new perspectives in relevant discussions, and a start to developing such perspectives has been offered here.

References

Aljughaiman, A., & Mowrer-Reynolds, E. (2005). Teachers' conceptions of creativity and creative students. *Journal of Creative Behavior, 39*, 17–34.

Baucus, M. S., Norton, W. I., Baucus, D. A., & Human, S. E. (2008). Fostering creativity and innovation without encouraging unethical behavior. *Journal of Organizational Behavior and Human Decision Processes, 72*, 117–135.

Bekoff, M., & Pierce, J. (2009). *Wild Justice: The Moral Lives of Animals.* Chicago, IL: University of Chicago Press.

Brandau, H., Daghofer, F., Hollerer, L., Kaschnitz, W., Kellner, K., Kirchmair, G., Krammer, L., & Schlagbauer. A. (2007). The relationship between creativity, teacher ratings on behaviour, age, and gender in pupils from seven to ten years. *Journal of Creative Behavior, 41*, 91–113.

Bruner, J. S. (1962). The conditions of creativity. In H. Gruber, G. Terrell, & M. Wertheimer (eds.), *Contemporary Approaches to Cognition* (pp. 1–30). New York: Atherton.

Chan, D. W., & Chan, L.-K. (1999). Implicit theories of creativity: Teachers' perception of student characteristics in Hong Kong. *Creativity Research Journal, 12*, 185–195.

Commonwealth of Australia (1999). *Higher Education Funding Report. 1999.* Canberra: Government Printer.

Cooper, C., Altman, W., & Garner, A. (2002). *Inventing for Business Success.* New York: Texere.

Cropley, A. J. (1997). Creativity: A bundle of paradoxes. *Gifted and Talented International, 12*, 8–14.

Cropley, A. J. (2009). Teachers' antipathy to creative students: Some implications for teacher training. *Baltic Journal of Psychology, 10*, 86–93.

Cropley, A. J. (2012). Creativity and education: An Australian perspective. *International Journal of Creativity and Problem Solving, 22*(1), 9–25.

Cropley, A. J., & Cropley, D. H. (2009). *Fostering Creativity: A Diagnostic Approach for Higher Education and Organizations.* Cresskill, NJ: Hampton Press.

Cropley, D. H., & Cropley, A. J. (2005). Engineering creativity: A systems concept of functional creativity. In J. C. Kaufman & J. Baer (eds.), *Faces of the Muse: How People Think, Work and Act Creatively in Diverse Domains* (pp. 169–185). Hillsdale, NJ: Lawrence Erlbaum Associates.

Cropley, D. H., & Cropley, A. J. (2013). *Creativity and Crime.* Cambridge: Cambridge University Press.

Cropley, D. H., Cropley, A. J., Kaufman, J. C., & Runco, M. A. (eds.) (2010). *The Dark Side of Creativity*. Cambridge: Cambridge University Press.

Csikszentmihalyi, M. (1996). *Creativity: Flow and the Psychology of Discovery and Invention*. New York: HarperCollins.

Dawson, V. L., D'Andrea, T., Affinito, R., & Westby, E. L. (1999). Predicting creative behavior: A reexamination of the divergence between traditional and teacher-defined concepts of creativity. *Creativity Research Journal, 12*, 57–66.

DeFillippi, R., Grabher, G., & Jones, C. (2007). Introduction to paradoxes of creativity: managerial and organizational challenges in the cultural economy. *Journal of Organizational Behavior, 28*, 511–521.

Fasko, D. (2000–2001). Education and creativity. *Creativity Research Journal, 13*, 317–328.

Feldhusen, J. F., & Treffinger, D. J. (1975). Teachers' attitudes and practices in teaching creativity and problem solving to economically disadvantaged and minority children. *Psychological Reports, 37*, 1161–1162.

Getzels, J. W., & Jackson, P. W. (1962). *Creativity and Intelligence*. New York: John Wiley & Sons.

Heimberg, C. L., Turk, D. S., & Mennin, R. G. (2004). *Generalized Anxiety Disorder: Advances in Research and Practice*. New York: Guilford Press.

Henning, B. G. (2005). *The Ethics of Creativity: Beauty, Morality, and Nature in a Processive Cosmos*. Pittsburgh, PA: University of Pittsburgh Press.

Holland, J. L. (1959). Some limitations of teacher ratings as predictors of creativity. *Journal of Educational Psychology, 50*, 219–223.

Karwowksi, M. (2007). Teachers' nominations of students' creativity: Should we believe them? Are the nominations valid? *The Social Sciences, 2*, 264–269.

Kim, K. H. (2011). The creativity crisis: The decrease in creative thinking scores on the Torrance Tests of Creative Thinking. *Creativity Research Journal, 23*, 285–295.

Makel, M. (2008). Help us creativity researchers, you're our only hope. *Psychology of Aesthetics, Creativity, and the Arts, 3*, 38–42.

Martin, M. W. (2006). Moral creativity. *International Journal of Applied Philosophy, 20*, 55–66.

Martin, R. L., & Florida, R. (2009). *Ontario in the Creative Age*. Toronto: Martin Prosperity Institute.

McWilliam, E., & Dawson, S. (2008). Teaching for creativity: Towards sustainable and replicable pedagogical practice. *Higher Education, 56*, 633–643.

Morgan, S., & Forster, J. (1999). Creativity in the classroom. *Gifted Education International, 14*, 29–43.

Mueller, J. S., Melwani, S., & Goncalo, J. A. (2012). The bias against creativity: Why people desire but reject creative ideas. *Psychological Science, 23*(1) 13–17.

Obuche, N. M. (1986). The ideal pupil as perceived by Nigerian (Igbo) teachers and Torrance's creative personality. *International Review of Education, 32*, 191–196.

Oral, G. (2006). Creativity of Turkish prospective teachers. *Creativity Research Journal, 18*, 65–73.

Oral, G., & Guncer, B. (1993). Relationship between creativity and nonconformity to school discipline as perceived by teachers of Turkish elementary school children, by controlling for their grade and sex. *Journal of Instructional Psychology, 20*, 208–214.

264   *The Ethics of Creativity*

Peterson, C., & Seligman, E. P. (2004). *Character Strengths and Virtues*. Oxford: Oxford University Press.

Pink, D. H. (2005). *A Whole New Mind: Moving from the Information Age to the Conceptual Age*. New York: Riverhead Books.

Plucker, J. A., Beghetto, R. A., & Dow, G. T. (2004). Why isn't creativity more important to educational psychologists? Potentials, pitfalls, and future directions in creativity research. *Educational Psychologist, 39*(2), 83–96.

Runco, M. A., & Jaeger, G. (2012). The standard definition of creativity. *Creativity Research Journal, 21*, 92–96.

Runco, M., & Johnson, D. (2002). Parents' and teachers' implicit theories of children's creativity. *Creativity Research Journal, 14*, 427–438.

Runco, M. A., Johnson, D. J., & Bear, P. K. (1993). Parents' and teachers' implicit theories of children's creativity. *Child Study Journal, 23*, 91–113.

Scott, C. L. (1999). Teachers' biases toward creative children. *Creativity Research Journal, 12*, 321–328.

Scott, G., Leritz, L. E., & Mumford, M. D. (2004). The effectiveness of creativity training: A quantitative review. *Creativity Research Journal, 16*, 361–388.

Smith, G. J. W., & Carlsson, I. (2006). Creativity under the Northern Lights: Perspectives from Scandinavia. In J. C. Kaufman & R. J. Sternberg (eds.), *International Handbook of Creativity* (pp. 202–234). New York: Cambridge University Press.

Staw, B. (1995). Why no-one really wants creativity. In C. Ford & D. Gioia (eds.), *Creative Action in Organizations: Ivory Tower Visions and Real Voices* (pp. 161–166). Thousand Oaks, CA: Sage.

Tan, A. G. (2003). Student teachers' perceptions of student behaviours for fostering creativity: A perspective on the academically low achievers. *Korean Journal of Thinking and Problem Solving, 13*, 59–71.

Tilbury, D., Reid, A., & Podger, D. (2003). *Action Research for University Staff: Changing Curricula and Graduate Skills towards Sustainability, Stage 1 Report*. Canberra: Environment Australia.

Torrance, E. P. (1959). *Explorations in Creative Thinking in the Early School Years: VIII. IQ and Creativity in School Achievement*. Minneapolis: Bureau of Educational Research, University of Minnesota.

Torrance, E. P. (1965). *The Minnesota Studies of Creative Thinking: Widening Horizons in Creativity*. New York: John Wiley & Sons.

West, M. A. (2002). Sparkling fountains or stagnant ponds: An integrative model of creativity and innovation implementation in work groups. *Applied Psychology, 51*, 355–387.

Westby, E. L., & Dawson, V. L. (1995). Creativity: Asset or burden in the classroom? *Creativity Research Journal, 8*, 1–10.

15
The Ethical Demands Made on Leaders of Creative Efforts

Michael Mumford, David R. Peterson, Alexandra E. MacDougall, and Thomas A. Zeni
University of Oklahoma, USA

and

Seana Moran
Clark University, USA

How should leaders make ethical decisions about the production and dissemination of creative products, when creative work adds to the normal leadership workload the complications of ill-defined tasks, uncertain solution paths, professionals who tend toward autonomy and exploration, and diversities of expertise that must collaborate? Ethics addresses the outcomes of decisions—typically the outcomes of decisions vis-à-vis others (Treviño, Brown, & Hartman, 2003). Creativity, the production of new ideas, and innovation, the translation of these ideas into viable new products and services (Mumford & Gustafson, 1988), are complex performances (Mumford, 2012) that can affect not only those involved in their production, but the organization and society more broadly.

Although many variables influence the success of creative efforts, leadership appears to be of special significance (Mumford et al., 2002). Creative people are especially sensitive to environmental context (Amabile et al., 1996; Oldham & Cummings, 1996); leaders define the environment people perceive (James, James, & Ashe, 1990). Creative work must be structured and evaluated because the ill-defined problems addressed often provide no structure or evaluation criteria (Lonergan, Scott, & Mumford, 2004; Marta, Leritz, & Mumford, 2005); leaders provide that scaffolding (Fleishman, 1953a, b). Creative work requires resources; leaders acquire those resources (Howell & Boies, 2004).

Of the many things leaders must do, insuring ethical conduct with respect to products, people, and processes is among the most important (Stenmark & Mumford, 2011). Although creative thought itself is a value-free phenomenon (Ludwig, 1998), creative solutions may or may not be ethical (Mumford et al., 2010). Leaders of creative efforts must take responsibility not only for the creative work being conducted, but also the ethical implications of this work. While leader ethics has received substantial attention (Bass & Steidlmeier, 1999; Brown & Treviño, 2006), few studies have examined the ethics of leading creative work (Ludwig, 1998; Mumford et al., 2010). Given the apparent impact of leadership on both the success of creative efforts and the ethical implications of creative work, we use a model of ethical decision-making to examine the critical activities needed to lead ethical creative ventures.

Leader ethics as sensemaking

The fundamental attribute for ethical leaders of creative efforts is sensemaking. Sonenshein's (2007) sensemaking theory states that ethical issues, like creative problem solutions, are inherently complex, ambiguous, and ill defined. Issues unfold over time as actions are taken, or not taken, with respect to the issue at hand. The leader's role is to formulate and articulate a framework for understanding and responding to the ethical issue. Leaders structure a dynamic, complex, unfolding set of operations to provide an environment for employees developing novel products to remain alert to the potential consequences of their work. The leader's challenges involve integrating various sources of information, forecasting, acquiring resources responsibly, and managing emotions.

Mumford and his colleagues have modeled the processes underlying sensemaking in leaders' ethical decision-making (Figure 15.1; Kligyte et al., 2008; Mumford et al., 2010; Thiel et al., 2012). This model assumes that professional guidelines frame ethical issues. However, the moral intensity of ethical issues also depends on how the leader perceives causes, effects of the situation on others, and requirements for resolving the issue. These guidelines and perceptions, along with leaders' past professional and personal experiences, help leaders frame the issue and activate emotions that could lead to engagement or withdrawal. The framing produces a coherent mental model, helping the group make sense of the situation in such a way that its members can forecast and evaluate possible actions, contingencies, and effects. Ongoing evaluation refines the mental model and helps leaders manage both the work tasks and group work processes to satisfy both

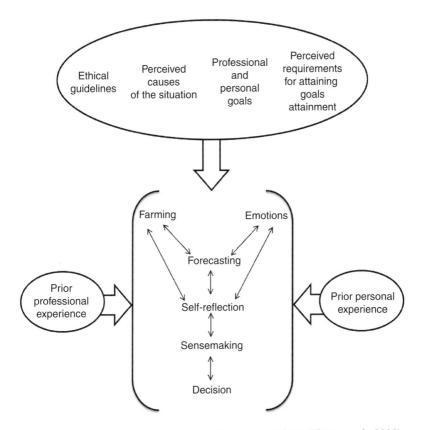

Figure 15.1 Sensemaking model of ethical decision-making (Kligyte et al., 2008).

creative and ethical goals. It is important that leaders self-reflect so that their past experiences do not introduce harmful biases into their leadership.

This model provides many insights into leadership generally. However, the focus here is on its implications for leaders of *creative* efforts. Leading creative groups involves *scanning the environment* for new themes that might be explored and exploited (Hughes, 1989; Verhaeghe & Kfir, 2002); *extensive planning*, since creative products, by virtue of their novelty, do not have specification sheets or established procedures (Buijis, 2008; Stockstrom and Herstatt, 2008); *defining a technical mission* by which the product can be evaluated, the development process can be monitored (Amabile & Kramer, 2011), and employees can receive feedback (Scott, Lonergan, & Mumford, 2005), which is often sought by creators to help guide their progress since creative production can be open-ended and nebulous (Farris, 1972); and *sharing mental models* with employees for the developing product (Day, Gronn, & Salas, 2006) as well as for work processes that reinforce creators' needs for intellectual challenge, mission clarity, support, prosocial exchange, and trust (Hunter, Bedell-Avers, & Mumford, 2007; Rego et al., 2007; Tierney, Farmer, & Graen, 1999; West & Sacramento, 2012). Furthermore, creative groups are embedded in larger organizations. Thus, leading creative endeavors also involves selling the value of the creative project to both senior executives (Howell & Boies, 2004) and other organizational subsystems needed to support the creative product (Jelinek & Schoonhoven, 1990). This salesmanship is not only to obtain required resources, but also to help others make sense of how creative projects may affect other organizational operations (Lassen, Waehrens, & Boer, 2009).

Creative efforts are higher in uncertainty than work projects with established procedures and goals. When the sensemaking model of leaders' ethical decision-making is applied to the creative leadership model (Robledo, Peterson, & Mumford, 2012), two key skills rise above the others: the leader's prowess in forecasting creative possibilities, contingencies, and their ethical consequences; and the leader's ability to integrate this forecast into a coherent story of which employees and other organizational stakeholders can make sense (Caughron et al., 2011; Stenmark et al., 2011). These two skills are not independent: failure to formulate a viable mental model hinders forecasting and backup planning; and planning, in turn, provides the basis for the creative mission that is the foundation for sensemaking. How do these two skills manifest in leading the work and work group, as well as leading within the organization?

Forecasting the work

Leaders must forecast the long-term implications of pursuing creative projects. Forecasting underpins planning, which involves the mental simulation of the consequences of future actions. Planning depends on the construction of viable mental models, forecasting the consequences of action based on these models, and forming action plans and backup plans (Mumford, Schultz, & Van Doorn, 2001; Xiao, Milgram, & Doyle, 1997). Groups and organizations typically hold leaders accountable for forecasting and backup planning. Without such work, leaders are criticized for being unprepared, having no idea what to do when "this event" occurred.

The exploratory nature of creative efforts provides leaders with a justification for limited forecasting and failure to formulate viable backup plans. However, forecasting and backup contribute not only to performance but also to effective exploitation of emergent opportunities (Patalano & Seifert, 1997). Forecasting improves when it is more extensive and unfolds over a longer timeframe (Byrne, Shipman, & Mumford, 2010; Shipman, Byrne, & Mumford, 2010). Thus, breaches in leadership ethics with respect to opportunity identification may arise from failure to envision how work will unfold over time and affect various constituencies.

Vigilance in scanning the environment for pertinent new information is important. Scanning is resource demanding, as evidenced in a common justification leaders make for ethical breaches: "I didn't have time to talk to them." Plus, people typically search for information that confirms their preexisting beliefs (Feist & Gorman, 1998). Leaders of creative efforts tend to focus on technical issues (Thiel et al., 2012). As a result, leaders often "do not hear" or "do not see" ethical issues broached by various sources searched, and do not frame decisions with respect to their ethical implications (Mumford et al., in press; Zuckerman, 1977). On the other hand, people tend to bias information search to superiors' expectations (Stenmark & Mumford, 2011). As a result, leaders may focus on organizational needs at the expense of technical issues pursued in creative work.

Given the unusually broad potential impact of creative efforts, both technical and organizational issues must be considered in the decision to pursue or not pursue a given creative opportunity, and leaders must be self-reflective in how they apply both. The point is aptly illustrated in Oppenheimer's reaction to the development of the atom bomb: we have created Vishnu the destroyer (Bird & Sherwin, 2005). Or, as McLaren (1993) notes:

> The man who invented the special cannon to defend his family ... was not less creative than Michelangelo, who only reluctantly created the sculpture of the Pope ... But a moral as well as artistic dilemma presented itself when it was decided to melt down that statue to build the cannon, and then fire it at the Pope! (p. 142)

Leaders, like people in general, are subject to optimistic bias (Hogarth, 1980). Limited forecasting may cause leaders to set goals that are unattainable, which demoralizes employees (Yukl, 2010), results in inappropriate expectations and premature investment (Einarsen, Aasland, &

Skogstad, 2007), and wastes resources (Shipman, Byrne, & Mumford, 2010). Furthermore, leaders discount obstacles that may arise in projects, which can sometimes stimulate more creativity to deal with the obstacle (Caughron & Mumford, 2008), but also can disrupt creative work (Alencar, 2012) and put undue pressure on employees' work process and resources.

Sensemaking of the work to the creative group

In defining the project mission, leaders must not only make sense of the project but must make the significance of the project clear to group members. This mission and the structuring of work vis-à-vis the mission influence leader performance and the success of creative efforts (Mumford et al., 2002; Friedrich et al., 2009). Failure to formulate a viable mental model will result in inadequate direction for the people asked to do the work (Day, Gronn, & Salas, 2006) and make it difficult for both leaders and the group to respond to crises that arise (Drazin, Glynn, & Kazanjian, 1999). Unfortunately, it is making the significance of the projects clear to team members where leaders often fail. The problem that arises is not sufficiently establishing a climate that incorporates both the needs of the employees and the needs of the project. These needs may conflict in tensions of boredom versus stimulation, and varying risk tolerances, competitiveness, production speed, and professional loyalties among group members (Hunter, Bedell-Avers, & Mumford, 2007). These tensions make creative work more demanding, and, perhaps, less likely to succeed. Yet, their interplay can also stimulate further creativity.

Relationships characterized by liking, trust, and performance typically contribute to positive leader–follower exchange (Graen & Uhl-Bien, 1998). Creative people, however, are not always likable (Feist, 1999). Creative projects are uncertain, leading to limited trust and overly close supervision (Andrews & Farris, 1972). Performance is not guaranteed on creative projects where the risk of failure is high (Paletz, 2012). Thus, the dilemma facing leaders of creative efforts is that the demands of creative work are inconsistent with the requirements for effective positive interpersonal exchanges.

Leaders sometimes do not take into account interpersonal interactions among team members and between a team, a leader, and the mission, which has a significant impact on creativity (Reiter-Palmon, Wigert, & de Vreede, 2012; Taggar, 2002). This failure to think about, protect, and manage a group's actions and reactions is common, in part as a result of the focus of creative leaders on the work per se. However, failure to forecast others' actions and interactions may make it impossible for leaders to encourage the focused collaboration required on most creative projects. Leaders must establish processes that allow autonomy as well as cohesion and collaboration (Greenberg, 1992; Sawyer, 2006), so that creative ideas can be freely expressed (Maier, 1950). Leaders must foster risk-taking, challenge followers, encourage followers to think, and insure positive interpersonal exchange to integrate an employee's sense of creative self-efficacy (Tierney & Farmer, 2010) with creative identity vis-à-vis the project (Jaussi, Randel, & Dionne, 2007).

Making sense of actions and reactions over time is particularly important with evaluative communications. Leaders of creative efforts tend to focus too much on technical issues that can demoralize employees (Gaddis, Connelly, & Mumford, 2004), and focus insufficiently on how the work process may change over the course of product development. Shifting contingencies refer to the tendency of leaders to punish people for failure to meet mission demands, yet mission demands can change as a creative product develops. Punishment is generally detrimental to creative work performance (Basadur & Basadur, 2011). Shifting standards become a problem when the leader imposes new, unexpected criteria during evaluation, often criteria not tied to the mission originally being pursued. The resulting inconsistency disrupts performance and motivation, inducing feelings of helplessness and loss of autonomy.

Because of the changing nature of creativity, timely monitoring of work is important to address emerging crises quickly (Drazin, Glynn, & Kazanjian, 1999). Leadership failures in monitoring include refusing to perceive the crisis and the need for action; not making sense of the crisis through effective causal analysis (Marcy & Mumford, 2007, 2010); and refusing to see the value of creative ideas for resolving the crisis simply because the ideas are unfamiliar (Licuanan, Dailey, & Mumford, 2007). Shared mental models may need to be changed to incorporate new creative ideas; indeed, the definition and adjustment of shared mental models are where leaders evidence their own creativity (Mumford, Connelly, & Gaddis, 2003).

Sensemaking of the work for the organization

Not only must leaders lead the work and the people doing the work, they must influence others in the organization supporting the work. This includes defining the role of the group's members within the wider organizational perspective. The first ethical problem that arises here is overly enthusiastic sales projections that create unrealistic expectations among project staff with regard to their contribution to the project and the value of the project for their career. The second ethical problem is poorly defined individual roles in the project and the contribution of the project to employees' career development. Leaders must define and sell creative projects to others beyond the group as well. Both fiscal and top management support are critical to the success of creative efforts (Dougherty & Hardy, 1996; Nohari & Gullati, 1996). Yet, because creative products involve novelty, senior managers and others in the organization may not have a good mental model for understanding the product's function or purpose. Leaders face three ethical issues: overselling, overresourcing, and integration with the rest of the organization.

First, to acquire support, leaders are often willing to promise that the creative effort will accomplish objectives it cannot meet, be completed in an unlikely timeframe, or involve less disruption of routine organizational operations than is actually likely to be the case. Leaders of creative efforts sell efficiently when they understand both the organization and the technology and help others understand both the promise of projects and the significance of the problems encountered (Kazanjian & Drazin, 2012). Yet, second, leaders must be realistic about their expectations. Creative projects often are chastised for going over budget or past deadlines because "unexpected issues arose." From an organizational perspective, resources invested in one effort cannot be invested elsewhere. Ethical leadership requires obtaining sufficient but not excessive resources to execute a project successfully, which returns to the leader's skill in providing a balanced appraisal of forecasted risks (Shipman, Byrne, & Mumford, 2010).

Finally, leaders of creative efforts must pay special attention to how their projects integrate with other organizational subsystems (Mumford et al., 2008). It becomes necessary, at some point, to involve these different subsystems in a creative project—the marketing people must be brought on board. Indeed, active support of creative efforts by various organizational subsystems is a crucial determinant of project success (Jelinek & Schoonhoven, 1990). Leaders must seek to understand and frame how supporting subsystems "see"—and are affected by—the creative effort (Kazanjian & Drazin, 2012). They must learn to "speak the language" of the expertise of the other subsystems that might be affected by the creative effort.

Most creative projects prove disruptive to the operations of other organizational systems. Accordingly, ethical leadership of creative efforts requires making potential disruptions clear to each supporting subsystem and taking action to minimize their effects. This disruption, however, is a two-way street. In turn, the leader must also manage the creative group to allow the other subsystems to provide needed support. The creative group cannot shun others with different expertise because the imported ideas or technologies were "not invented here" (Ancona & Caldwell, 1992). Despite creative groups sometimes having trouble cohering, if a creative group is highly cohesive it may dismiss ideas about capabilities from outside the team (Gerstenberger & Allen, 1968; Keller, 1989). Leaders must engage in ongoing, inclusion-building sensemaking of the multiple contributions that varied expertise groups bring to the creative effort organization-wide.

Conclusion

We crossed a model of ethical decision-making (for example, Kligyte et al., 2008; Mumford et al., 2008; Thiel et al., 2011) with a model for leading creative efforts (Mumford, Connelly, & Gaddis, 2003; Robledo, Peterson, & Mumford, 2012) to show that ethical leadership of creative efforts can be understood in terms of sensemaking (Sonenshein, 2007). In creative work, it is particularly important that this sensemaking extends beyond what is observable now to include forecasting what could happen in the future. Sensemaking as a model for ethical decision-making by leaders of creative efforts results in some good news, some bad news, and some ambiguous news. The good news: Leading the *work* ethically is generally consistent with effective work practices for the leaders of creative efforts (Mumford et al., in press). Ethics and creativity support each other. The bad news: Leading the *group* often requires sacrificing at least some aspect of the work. Risk-taking must be encouraged to establish a viable climate for creative work, even though these risks are not guaranteed to produce a viable creative product and create a more turbulent work environment. Creativity and ethics requirements may diverge, and leaders must make extra efforts to help employees make sense of the situation in a productive way. The ambiguous news: Leading within the *organization* ethically requires forecasting how organizational subsystems understand the needs and implications of a creative effort. These needs and implications may cooperate or conflict, thus creativity and ethics requirements may converge or diverge. In summary, the needs of the people involved may not be consistent with the needs of the work. For example, challenging employees may not be consistent with positive interpersonal exchange. The conflict between ethical leadership of the group or organization and ethical leadership of the work suggests that extensive training in sensemaking and social skills is required of leaders. Failure to invest in this training suggests why leaders who can "produce the goods" are often not thought well of by the people they are leading in doing this work.

References

Alencar, E. M. L. S. (2012). Creativity in organizations: Facilitators and inhibitors. In M. Mumford (ed.), *Handbook of Organizational Creativity* (pp. 483–515). London: Elsevier.

Amabile, T., & Kramer, K. (2011). *The Progress Principle*. Boston, MA: Harvard Business Review Press.

Amabile, T. M., Conti, R., Coon, H., Lazenby, J., & Herron, M. (1996). Assessing the work environment for creativity. *Academy of Management Journal, 39,* 1154–1184.

Ancona, D., & Caldwell, D. (1992). Demography and design: Predictors of new product team performance. *Organization Science, 3,* 321–341.

Andrews, F. M., & Farris, G. F. (1972). Time pressure and the performance of scientists and engineers: A five year panel study. *Organizational Behavior and Human Performance, 8,* 185–200.

Basadur, M., & Basadur, T. (2011). Where are the generators? *Psychology of Aesthetics, Creativity, and the Arts, 5,* 29–42.

Bass, B. M., & Steidlmeier, P. (1999). Ethics, character, and authentic transformational leadership behavior. *Leadership Quarterly, 10,* 181–218.

Bird, K., & Sherwin, M. J. (2005). *American Prometheus: The Triumph and Tragedy of J. Robert Oppenheimer.* New York: Vintage Books.

Brown, M. E., & Treviño, L. K. (2006). Ethical leadership: A review and future directions. *Leadership Quarterly, 17,* 595–616.

Buijis, J. (2008). Action planning for new product development projects. *Creativity and Innovation Management, 17,* 319–333.

Byrne, C. L., Shipman, A. S., & Mumford, M. D. (2010). The effects of forecasting on creative problem-solving: An experimental study. *Creativity Research Journal, 22,* 119–138.

Caughron, J. J., & Mumford, M. D. (2008). Project planning: The effects of using formal planning techniques on creative problem-solving. *Creativity and Innovation Management, 17,* 204–215.

Caughron, J. J., Antes, A. L., Stenmark, C. K., Thiel, C. E., Wang, X., & Mumford, M. D. (2011). Sensemaking strategies for ethical decision making. *Ethics and Behavior, 21*(5), 351–366.

Day, D. V., Gronn, P., & Salas, E. (2006). Leadership in team-based organizations: On the threshold of a new era. *Leadership Quarterly, 17,* 211–216.

Dougherty, D., & Hardy, C. (1996). Sustained product innovation in large, mature organizations: Overcoming innovation-to-organization problems. *Academy of Management Journal, 39,* 1120–1153.

Drazin, R., Glynn, M. A., & Kazanjian, R. K. (1999). Multilevel theorizing about creativity in organizations: A sensemaking perspective. *Academy of Management Review, 24,* 286–329.

Einarsen, S., Aasland, M. S., & Skogstad, A. (2007). Destructive leadership behavior: A definition and conceptual model. *Leadership Quarterly, 18,* 207–216.

Farris, G. F. (1972). The effect of individual role on performance in innovative groups. *R&D Management, 3*, 23–28.

Feist, G. J. (1999). The influence of personality on artistic and scientific creativity. In R. Sternberg (ed.), *Handbook of Creativity* (pp. 273–296). New York: Cambridge University Press.

Feist, G. J., & Gorman, M. E. (1998). The psychology of science: Review and integration of a nascent discipline. *Review of General Psychology, 2*, 3–47.

Fleishman, E. A. (1953a). Leadership climate, human relations training, and supervisory behavior. *Personnel Psychology, 6*, 205–222.

Fleishman, E. A. (1953b). The description of supervisory behavior. *Journal of Applied Psychology, 37*, 1–6.

Friedrich, T., Vessey, W., Schuelke, M., Ruark, G., & Mumford, M. (2009). A framework for understanding collective leadership: The selective utilization of leader and team expertise within networks. *Leadership Quarterly, 20*, 933–958.

Gaddis, B., Connelly, S., & Mumford, M. D. (2004). Failure feedback as an affective event: Influences of leader affect on subordinate attitudes and performance. *Leadership Quarterly, 15*, 663–686.

Gerstenberger, P. C., & Allen, T. J. (1968). Criteria used by research and development engineers in the selection of an information source. *Journal of Applied Psychology, 52*, 272–279.

Graen, G. B., & Uhl-Bien, M. (1998). Relationship-based approach to leadership: Development of leader-member exchange (LMX) theory of leadership over 25 years: Applying a multi-level multi-domain perspective. In F. Dansereau & F. J. Yammarino (eds.), *Leadership: The Multi-Level Approaches* (pp. 103–134). Stamford, CT: JAI Press.

Greenberg, E. (1992). Creativity, autonomy, and the evaluation of creative work: Artistic workers in organizations. *Journal of Creative Behavior, 26*, 75–80.

Hogarth, R. M. (1980). *Judgment and Choice: The Psychology of Decision.* Chichester, UK: John Wiley & Sons.

Howell, J. M., & Boies, K. (2004). Champions of technological innovation: The influences of contextual knowledge, role orientation, idea generation, and idea promotion on champion emergence. *Leadership Quarterly, 15*, 130–149.

Hughes, T. P. (1989). *American Genesis: A History of the American Genius for Invention.* New York: Penguin.

Hunter, S. T., Bedell-Avers, K. E., & Mumford, M. D. (2007). Climate for creativity: A quantitative review. *Creativity Research Journal, 19*, 69–90.

James, L. R., James. L. A., & Ashe, D. K. (1990). The meaning of organizations: The role of cognition and values. In B. Schneider (ed.), *Organizational Climate and Culture* (pp. 40–84). San Francisco, CA: Jossey-Bass.

Jaussi, K., Randel, A., & Dionne, S. (2007). I am, I think I can, and I do: The role of personal identity, self-efficacy, and cross-application of experiences in creativity at work. *Creativity Research Journal, 19*, 247–258.

Jelinek, M., & Schoonhoven, C. B. (1990). *The Innovation Marathon: Lessons Learned from High Technology Firms.* Oxford: Blackwell.

Kazanjian, R. K. & Drazin, R. (2012). Organizational learning, knowledge management, and creativity. In M. D. Mumford (ed.) *Handbook of Organizational Creativity* (pp. 547–568). London: Elsevier.

Keller, R. T. (1989). A test of the path-goal theory of leadership with need for clarity as a moderator in research and development organizations. *Journal of Applied Psychology, 74*, 208–212.

Kligyte, V., Marcy, R. T., Waples, E. P., Sevier, S. T., Godfrey, E. S., Mumford, M. D., & Hougen, D. F. (2008). Application of a sensemaking approach to ethics training for physical sciences and engineering. *Science and Engineering Ethics, 14,* 251–278.

Lassen, A., Waehrens, B., & Boer, H. (2009). Re-orienting the corporate entrepreneurial journey: Exploring the role of middle management. *Creativity and Innovation Management, 18,* 16–23.

Licuanan, B. F., Dailey, L. R., & Mumford, M. D. (2007). Idea evaluation: Error in evaluating highly original ideas. *Journal of Creative Behavior, 41,* 1–27.

Lonergan, D. C., Scott, G. M., & Mumford, M. D. (2004). Evaluative aspects of creative thought: Effects of idea appraisal and revision standards. *Creativity Research Journal, 16,* 231–246.

Ludwig, A. M. (1998). Method and madness in the arts and sciences. *Creativity Research Journal, 11,* 93–101.

Maier, N. R. (1950). The quality of group discussions as influenced by the discussion leader. *Human Relations, 3,* 155–174.

Marcy, R. T., & Mumford, M. D. (2007). Social innovation: Enhancing creative performance through causal analysis. *Creativity Research Journal, 19,* 123–140.

Marcy, R. T., & Mumford, M. D. (2010). Leader cognition: Improving leader performance through causal analysis. *Leadership Quarterly, 21,* 1–19.

Marta, S., Leritz, L. E., & Mumford, M. D. (2005). Leadership skills and group performance: Situational demands, behavioral requirements, and planning. *Leadership Quarterly, 16,* 97–120.

McLaren, R. B. (1993). The dark side of creativity. *Creativity Research Journal, 6*(1–2), 137–144.

Mumford, M. D. (ed.) (2012). *Handbook of Organizational Creativity.* London: Elsevier.

Mumford, M. D., & Gustafson, S. B. (1988). Creativity syndrome: Integration, application, and innovation. *Psychological Bulletin, 103,* 27–43.

Mumford, M. D., Connelly, S., & Gaddis, B. (2003). How creative leaders think: Experimental findings and cases. *Leadership Quarterly, 14,* 411–432.

Mumford, M. D., Schultz, R. A, & Van Dorn, J. R. (2001). Performance in planning: Processes, requirements, and errors. *Review of General Psychology, 5,* 213–240.

Mumford, M. D., Connelly, S., Brown, R. P., Murphy, S. T., Hill, J. H., Antes, A L., Waples, E. P., & Devenport, L. D. (2008). A sensemaking approach to ethics training for scientists: Preliminary evidence of training effectiveness. *Ethics & Behavior, 18,* 315–339.

Mumford, M. D., Gibson, C., Giorgini, V., & Mecca, J. (in press). Leading for creativity: People, processes, and systems. *Oxford Handbook of Leadership and Organizations.*

Mumford, M. D., Scott, G. M., Gaddis, B. H., & Strange, J. M. (2002). Leading creative people: Orchestrating expertise and relationships. *Leadership Quarterly, 13,* 705–750.

Mumford, M. D., Waples, E. P., Antes, A. L., Brown, R. P., Connelly, S., Murphy, S. T., Devenport, L. D. (2010). Creativity and ethics: The relationship of creative and ethical problem-solving. *Creativity Research Journal, 22,* 74–89.

Nohari, K., & Gulati, D. (1996). Is slack good or bad for innovation? *Academy of Management Journal, 39,* 799–825.

Oldham, G. R., & Cummings, A. (1996). Employee creativity: Personal and contextual factors at work. *Academy of Management Journal, 39,* 607–634.

Paletz, S. B. F. (2012). Project management of innovative teams. In M. Mumford (ed.), *Handbook of Organizational Creativity* (pp. 483–515). London: Elsevier.

Patalano, A. L., & Seifert, C, M. (1997). Opportunistic planning: Being reminded of pending goals. *Cognitive Psychology, 34,* 1–36.

Rego, A., Sousa, F., Pina e Cunha, M., Correia, A., and Saur-Amaral, I. (2007). Leader self-reported emotional intelligence and perceived employee creativity: An exploratory study. *Creativity and Innovation Management, 16,* 250–264.

Reiter-Palmon, R., Wigert, B., & de Vreede, T. (2012). Team creativity and innovation: The effect of group composition, social processes, and cognition. In M. Mumford (ed.), *Handbook of Organizational Creativity* (pp. 295–326). London: Elsevier.

Robledo, I. C., Peterson, D. R., & Mumford, M. D. (2012). Leadership of scientists and engineers: A three-vector model. *Journal of Organizational Behavior, 33,* 140–147.

Sawyer, R. K. (2006). *Explaining Creativity: The Science of Human Innovation.* New York: Oxford University Press.

Scott, G. M., Lonergan, D. C., & Mumford, M. D. (2005). Contractual combination: Alternative knowledge structures, alternative heuristics. *Creativity Research Journal, 17,* 21–36.

Shipman, A. S., Byrne, C. L., & Mumford, M. D. (2010). Leader vision formation and forecasting: The effects of forecasting extent, resources, and timeframe. *Leadership Quarterly, 21,* 439–456.

Sonenshein, S. (2007). The role of construction, intuition, and justification in responding to ethical issues at work: The sensemaking-intuition model. *Academy of Management Review, 32,* 223–239.

Stenmark, C. K., & Mumford, M. D. (2011). Situational impacts on leader ethical decision-making. *Leadership Quarterly, 22,* 942–955.

Stenmark, C. K., Antes, A. L., Thiel, C. E., Caughron, J. J., Wang, X., & Mumford, M. S. (2011). Consequences identification in forecasting and ethical decision-making. *Journal of Empirical Research on Human Research Ethics, 6,* 25–32.

Stockstrom, C., & Herstatt, C. (2008). Planning and uncertainty in new product development. *R&D Management, 38,* 480–490.

Taggar, S. (2002) Individual creativity and group ability to utilize individual creative resources: A multilevel model. *Academy of Management Journal, 45,* 315–330.

Thiel, C. E., Bagdasarov, Z., Harkrider, L., Johnson, J. F., & Mumford, M. D. (2012). Leader ethical decision-making in organizations: Strategies for sensemaking. *Journal of Business Ethics, 107,* 49–64.

Thiel, C. E., Connelly, S., Harkrider, L., Devenport, L. D., Bagdasarov, Z., Johnson, J. F., & Mumford, M. D. (2011). Case-based knowledge and ethics education: Improving learning and transfer through emotionally rich cases. *Science and Engineering Ethics.* Advance online publication.

Tierney, P., & Farmer, S. M. (2010). Creative self-efficacy development and creative performance over time. *Journal of Applied Psychology, 96,* 277–293.

Tierney, P., Farmer, S. M., & Graen, G. B. (1999). An examination of leadership and employee creativity: The relevance of traits and relationships. *Personnel Psychology, 52,* 591–620.

Treviño, L. K., Brown, M., & Hartman, L. P. (2003). A qualitative investigation of perceived executive ethical leadership: Perceptions from inside and outside the executive suite. *Human Relations, 55*, 5–37.

Verhaeghe, A., & Kfir, R. (2002). Managing innovation in a knowledge intensive technology organization (KITO). *R&D Management, 32*, 409–417.

West, M. A., & Sacramento, C. A. (2012). Creativity and innovation: The role of team and organizational climate. In M. Mumford (ed.), *Handbook of Organizational Creativity* (pp. 483–515). London: Elsevier.

Xiao, Y., Milgram, P., & Doyle, D. J. (1997). Capturing and modeling planning expertise in anesthesiology: Results of a field study. In C. E. Zsambok & G. Klein (eds.), *Naturalistic Decision Making* (pp. 197–205). Hillsdale, NJ: Lawrence Erlbaum Associates.

Yukl, G. (2010). *Leadership in Organizations*. Upper Saddle River, NJ: Prentice-Hall.

Zuckerman, H. (1977). *Scientific Elite: Nobel Laureates in the United States*. New York: Free Press.

Part IV
Horizons

An Ethics of Possibility

Seana Moran
Clark University, USA

Why and how do people intentionally choose to add variation and diversity to society and culture, rather than mimic others' successes or otherwise support the status quo? These creative acts, and the ideas, products, or procedures they spawn, also create a problem space—or what I believe *should* be a problem space—at the intersection of creativity and ethics. Most of the scholars in this volume start from the perspective of creativity and examine how morality plays into creative endeavors. In this chapter, I provide the alternative starting point of moral reasoning and moral development and examine what effect creativity may have on ethics. How does injecting creativity—novel and useful ideas or products that can transform the cultures into which they are launched (Csikszentmihalyi, 1988; see also Moran, 2010d)—affect notions of what people consider to be good?

One assumption in moral development is that we learn the mores and ethics that steer our lives through interaction with others from a very young age, such as through Vygotsky's (1978) "proximal zone of development" (see Tappan, 1997). These moral tenets tend to become assumptions, tacit knowledge we live by (Polanyi, 1958) that often seems intuitive (Haidt, 2001) but, when prompted or challenged, may be justified through reasoning (Haidt & Graham, 2007). Since mores and ethics are "lenses we see by," like eyeglasses we wear, we may not notice them when life is going relatively smoothly. We mindlessly use our ethics while going about our daily activities. It is primarily when we encounter an ethical dilemma or a foreign ethical viewpoint that challenges our ethical framework that we—often tentatively or reluctantly—look more closely at the ways in which we consider what we *ought* to do.

The trolley problem meets creative problem-solving

Here is the situation: You are supervising a crew working on railroad tracks. You are on a bridge overlooking your crew, who work about a quarter mile away. You turn around and see a runaway train barreling down the track. You have two options. One, do nothing and the train will kill your crew, all five of them. Two, flip a switch on the bridge next to you, the train will divert to a parallel track, and only one person will be killed. What do you do? A variant is that your second option is to push another supervisor, standing next to you, off the bridge onto the track to stop the train before it hits the crew. What do you do?

In moral philosophy and psychological moral reasoning studies, this is a variant of a classic moral dilemma called the trolley problem (Thomson, 1976). Moral reasoning entails determining what you or I *ought* to do in a situation (normative ethics, which prescribes ideals; Aristotle, 1999; Rawls, 1971, 2001) as well as what people *actually* do in a situation (descriptive ethics, which usually outlines our psychological biases and errors that inhibit us from reaching the normative ideal; Gigerenzer, 2010; Kohlberg, 1981; Rest, 1986). The point is: What is our duty or obligation in the presented situation? Researchers aim to make visible individuals' tacit frames for approaching such ethical quandaries.

People differ in the way they make distinctions and judgments about right and wrong, benefits and harms, good and evil. Although different communities and cultures set forth ideals for such perceptions, decisions, and behaviors (Haste & Abrahams, 2008; Rozin et al., 1999; Shweder, Mahapatra, & Miller, 1987), individuals vary in the distinctions and judgments they make. Several theories have been put forth to categorize individual differences based on criteria for evaluating the situation, perhaps the best known being Kohlberg's (1981) stage theory of moral development, and the many subsidiary theories others built from it (for example, Gilligan, 1982; Rest, 1986). Considerable controversy remains regarding whether the moral frames espoused by these theories are stages or types (Krebs & Denton, 2005) and whether moral reasoning translates into moral action (Blasi, 1980). Still, many of the frames can be readily adapted to a general framework based on the perspectives the person considers in his or her moral reasoning. The first perspective focuses on *me* as the key criterion for ethics: fight for a personal gain or avoid punishment by an external authority. The second perspective focuses on *relationships* as the key criterion: live up to your close relations', peer group's, or community's expectations of you, or keep the peace by obeying laws. Mimic what the majority of people abide by;

watch and learn from others as your moral guides. Most studies show that this stage is where most adults fall (Edwards, 1980). The third perspective focuses on *principles* as the key criterion, either based on the notion of human rights and the duties and social contracts needed to enforce those rights, or on even more abstract principles of justice or caring or sustainability (see Gibbs, 2003).

People often do not clearly articulate their perspectives or the underlying criteria. For the most part, ethics and morality are considered stable backgrounds to our lives. When was the last time we had an amendment to the Ten Commandments? How often do corporations revisit their ethics codes? Usually, these addenda arise when an institution or community experiences a shock, a disturbance, a perturbation to the status quo.

Creativity can be one of those shocks. The introduction and propagation of novel ideas or products through a culture can bring ethics to the foreground. Of course, most novel introductions are adaptations—small movements beyond what already exists in the culture, often through transfers across conceptual categories, or minor tweaks and extensions on current ideas or products (Baughman & Mumford, 1995), which sometimes are so close to existing options they are not even viewed as creative (Kirton, 1976; Mumford & Gustafson, 1988).

Adaptations may rock cultures and ethics, but not capsize them. A case in point is the myriad of design competitions held at universities like Harvard or MIT or Stanford, in which student teams build prototypes of new products and technologies. What students come up with is very impressive, but their products tend to be adaptations, transfers of properties of one type of product to another: bicycles' power-assist technology transferred to wheelchairs, vacuum packaging transferred to suitcases, vending machines transferred to bike helmet distribution. These adaptations can be very successful because people can understand them fairly quickly (Mumford & Gustafson, 1988). They are not so different that they cause confusion or distress, but different enough to stimulate interest. Engineers are masters of this adaptive process—of making improvements while staying within the constraints of reality and utility (see D. Cropley, this volume).

Perhaps it is, in part, because most novelties are adaptations of what already exists that creativity sometimes is assumed to be a force for good (Florida, 2002; Guilford, 1950; Maslow, 1973; McLaren, 1993; Moran, 2010d). New *is* improved. Creative work is taken for granted as something valuable to be supported by cultural resources: think of all the advancements in the last few centuries that different societies have made based on the intentional introduction and propagation of novel and useful ideas. I believe this positive view of creativity is part of the reason much scholarship has explored the cognitive and motivational processes *to come up with* new ideas, whereas less scholarship has addressed how people determine *which ideas are worthy* to keep, especially in light of their ethical ramifications.

But there is another category of creativity: innovation (Feldman, Csikszentmihalyi, & Gardner, 1994; Kirton, 1976), also thought of as the other end of the spectrum of creativity. Innovation is rarer than adaptation, but when it occurs, it transforms part or all of the culture. Often referred to as historical, eminent, or "Big-C" creativity (see Feldman, Csikszentmihalyi, & Gardner, 1994; Runco & Richards, 1989), innovations are shocks—like a cliff, which leads to waterfalls that change the landscape. An innovation in a culture or a social system surprises (Bruner, 1962) and redirects the group's thinking or behavior. Innovation takes some dimension of a problem or situation, which previously was an assumption, taken for granted and unspoken, and turns it into a variable. It allows that dimension to take on different values—multiple possible answers rather than one agreed-on right answer. It introduces variation where before there was none. In art, for example, Picasso took three-dimensional perspective and, through transparency, squashed the perspective into two dimensions. Jackson Pollock varied the viscosity or resistance of paint. At the time, most artists considered these tactics to be "error"—why would anyone want to do that? They were the *wrong* way to paint. But enough other artists and art patrons were intrigued by the results that these innovations stuck. These new techniques became *right*.

With innovation, uncertainty (Kagan, 2009), turbulence (Hall & Rosson, 2006), disruption (Christensen, Craig, & Hart, 2001; Florida, 2012), and destruction (Schumpeter, 1939) may ensue. No wonder many people are ambivalent about creativity (see Moran, 2009a, 2010a, b, c, d): it sounds like a nightmare of upheaval to the users and bystanders on whom the novelty is thrust. This upheaval may be fast or long-lasting, depending on how quickly or slowly the innovation is adopted. Individuals have different levels of comfort with novelty and adopt novelties on different timeframes (Moran, 2010c; Rogers, 1983[1962]). Once the novelty becomes the new norm or standard, the upheaval settles down. But until then, novel introductions can affect identity, status, prestige, allocation of resources, and so on. They make—and break—careers and institutions. Consider all the physicists who dedicated their lives to the "ether problem" when Einstein's theory of relativity came along and made the concept of ether irrelevant; or, more recently, the obsolescence of telegraphs with the introduction of telephones, horse-drawn buggies because of cars, and human "calculators" as a job category when computers emerged.

These shocks of creativity not only can shift field norms or accepted canons of knowledge, they also can disturb ethical foundations. Transformative creative works, at their most powerful, can unsettle moral certitudes by presenting new perspectives and possibilities that are unnerving. Copernicus's and Galileo's work in astronomy made people consider that humans may not be the center of everything. Darwin's work—and, increasingly, the discovery of many planets in the universe—destabilizes our belief that humans on earth are the pinnacle of life's achievement in the universe. Several recent technological innovations have brought with them ethical quandaries and ramifications that many of us would rather do without: the atomic bomb and nuclear energy, life-support systems and the vegetative state, "mashup music" and other bricolage techniques that question the notion of ownership. Some people might consider these new works socially and morally irresponsible. After all, as little as 50 years ago, jazz music was considered deviant, irresponsible, and threatening to the moral fiber of American society (Becker, 1963; Lopes, 2005; Stebbins, 1966).

Creativity entices us—or forces us—to consider ethical dimensions and options that were previously hidden or dormant. Are we killers if we employ drones or assisted suicide methods? Are social media technologies improving or diminishing our capabilities to relate to each other, empathize, communicate? Will genetically modified foods make us healthier or turn us into hybrid mutants if we consume their Frankenstein-like combination of various species' DNA? When faced with these new dilemmas, how can we devise effective ways to tackle them? Although these issues show us *reacting* to creativity thrust on us, we also have the opportunity to use creativity *proactively* to address ethics (Kaufmann, 2004).

Expanding the scope of ethics

Let us return to the moral dilemma of the trolley problem. As originally designed, presumably all of the available options are described within the dilemma. We must choose one or the other of the options presented to us. What is less often discussed in the moral reasoning, ethics, and moral development literatures is that the dilemma does not give us a complete picture of possibilities (see Cua, 1978; Weston,

2006). The dilemma narrowly scopes the situation for us, like a multiple-choice test question with no "other, please specify" option. We are primed *not* to consider that the situation occurs within a wider context, yet the situation may contain many elements not specified. Using many of the tools specified in the creativity literature, including perception (Wertheimer, 1945), conceptual combination (Scott, Lonergan, & Mumford, 2005), and adaptation (Kirton, 1976), perhaps we can apply an *ethics of possibility*, an ethics built not on rules but on opportunities.

Let us try it: What are other possibilities for the trolley problem besides doing nothing, flipping a switch, and pushing the other supervisor off the bridge? When I pose this question to students or presentation attendees, here are a few responses I hear:

- I could jump off the bridge and let my body stop the train.
- I could blow up the bridge I'm on and derail the train.
- I could lasso the train as it comes under the bridge using some type of net draped from the bridge.
- I might have a cell phone and could call the crew foreman to get out of the way.
- I might be relieved because I was going to have to fire my crew that afternoon anyway, and now I won't have to go through that agony if the train hits them.
- Perhaps it would've occurred to me, since I've been a supervisor for many years, that at some point a runaway train might be a possibility, and had set up a contingency plan with the foreman down below to "raise a red flag" in such a situation.

With each of these other options, we take some dimension of the situation, which in the original dilemma was taken for granted and unspoken, and we turn the dimension into something that can take on different values. For example, we varied who is sacrificed, availability of tools, communication options, anticipation of upcoming events, and history with team dynamics. Some of the options presented may have frightened or even disgusted us by bumping into our ethical limits (Haidt, Koller, & Dias, 1993; Rozin et al., 1999)—most likely the option of suicide or allowing the crew to die for our own benefit. These nearly instantaneous, intuitive, emotional triggers tied to our morals can enhance or diminish our capacity for creativity in our ethical thinking.

This expansion the trolley problem is an example of how transformative creativity can make societies and their ethics "wobble" by exposing previously unspoken assumptions, producing variations and alternatives where we thought right and wrong were "all settled," introducing misalignments between "oughts" and "dos," and, perhaps most of all, increasing uncertainty. Creativity takes what is considered "right" or "good," which people prefer to be a fixed point, and turns it into a moving target. This change can breed anxiety (Jaques, 1955). How can we more effectively think about an ethics that does not intuitively trigger moral disgust or anxiety when a new possibility is introduced?

Broadening our ethical horizons

Possibilities are shadows that actuality casts (see Cremin, Burnard, & Craft, 2006). When the light of thinking shifts, so do possibilities. Possibilities are difficult to chase, even harder to catch. But they tease us with a likelihood, a capability of coming into actuality—like Pinocchio, of "becoming real"—and bringing with them the potential for interesting adventures. Without possibilities, we live in a world of close-knit, often redundant information (Granovetter, 1983). We feel confident that we know what we need to know, but it is a false confidence blind to what we refuse to see. This conundrum is where we find ourselves today—enticed by, yet fearing, the what-not-yet-is. What is the right thing to do?

Creativity is strongly valued by political and business leaders, educational institutions, and consumers. New ideas and products are expected to improve productivity, personal wellbeing, and professional achievement. The belief is that, without new ideas, we will lose our competitive advantage, stifle economic growth, become bored (Moran, 2010c). Many governments, industry organizations, and scholars plan for "creative industries" (European Commission, 2009; Pratt & Jeffcutt, 2009), "creative cities" (Cho, 2010; Florida, 2002), "innovation labs" (Magedley & Birdi, 2009), and "innovation networks" (Tuomi, 2001) to help stimulate creativity among more individuals.

Yet, creativity also triggers conflict and struggle between traditional and emerging ways of life and institutions. Many worry that creativity breeds exploitation, inequality, waste of resources, unnecessary high risk, and corruption (Cho, 2010; Oakley, 2004; Ryder, 2011). Financial innovations launched into the mainstream market in the 2000s wreaked havoc worldwide (Dudley, 2010). Cell phones and other electronics may provide new avenues to deliver healthcare (Patrick et al., 2008), but also create physical and psychological health effects (David-Ferdon & Hertz, 2007; Volkow et al., 2011). Both privacy and ownership of ideas have become controversial with social media and internet search (Darnton, 2009; Liu et al., 2011). But creativity's influence is not an either–or situation. We have other options beyond unbridled optimism and stifling fear.

Bruner's "beyond the given" applied to ethics

An ethics of possibility goes "beyond the given" (Bruner, 1962) of rule-based ethics. It provides a valuing of what is beyond the known and accepted, a humility about what we already understand against the vastness of what we may not, and a courage to pursue unknowns. The world is more wondrous than what already exists from the fruits that past creative minds have given us. Perhaps there *is* a moral imperative to create (Gruber, 1993), to do our part in contributing to culture as it proceeds over time. My students and I talk about three primary tools for thinking ethically "beyond the given" (see also Moran, 2010c):

- *Contribution*: Each of us must do our part in the bigger picture of life's jigsaw puzzle. Learning what the culture already has to offer is important as a launchpad for our own contributions, but we should not make carbon-copy contributions of what others have already done because we have different possibilities open to us.
- *Consideration*: Each of us matters, in the big picture, but it is not "all about me" or about strictly self-interested personal gains. What we do matters to both known and unknown others. Our contributions can stay local to our friends or family, or "go viral" to more widely influence people we do not know personally. Regardless of whether we meet these unknown others, because we are part of neighborhoods, institutions, and other social webs, our decisions and actions can propagate further than we expect. We are "all in this together" even if we are not conforming to the same goal or way of life.
- *Kaizen*: Traditionally, kaizen is a Japanese method of continuous improvement by tracing the symptoms of a problem back to an original cause. The classic example is the problem of irregular ink coverage on a printed sheet of paper, which is determined—through a process of continually asking of each suggested solution/cause: "And what caused *that*?"—to be caused by a loose bolt in the printing press. What would happen if kaizen were applied not to look backward in time to causes, but to look forward in time to the implications of an action, to a purposeful consequence? Kaizen would entail asking of ourselves, our contributions, and especially our creative works: What effect is it possible and probable to have, not merely during our lifetimes but as a legacy?

Beyond foresight toward anticipation

Creativity is a temporally dependent act (Csikszentmihalyi & Nakamura, 2007; Moran, 2010c, d). Coming up with something new and useful takes time. Judging something new takes time. Accepting, building social infrastructure for, and propagating something new take time. Some think of creativity not as moving forward into the future, but as inventing the future itself. Our ethics, then, must not be limited to the here-and-now.

A new idea can be difficult to perceive, conceive, and control (Kagan, 2009). The most transformative creativity can have ripple effects for decades. Einstein's relativity, a theory in physics, spawned innovations in literary forms (Whitworth, 2002). Darwin's evolution by natural selection, a theory in biology, underpins business strategies (Marks, 2002). Picasso's and Braque's Cubism, a style of painting, influenced engineering education (Mina, Kemmet, & Gerdes, 2007). It is unclear who—if anyone—will reap the benefits of a creative idea or product, even a successful one (Simonton, 1994, 2004; Tenner, 1996). It is also unclear whether—or when—costs of past innovations will come due, such as pollution from the industrial age, or urban congestion from cars, or obesity from processed foods (Tenner, 1996).

Several studies have examined the role of foresight or forecasting in both creativity and ethics (Runco & Nemiro, 2003). Thinking through the possible implications of our plans goes hand in hand with the quality and originality of what we end up producing (Byrne, Shipman, & Mumford, 2010; Stenmark et al., 2010). Forecasting is easier in a stable environment than in a discontinuous environment resulting from a creative transformation (Welch, Bretschneider, & Rohrbaugh, 1998). The future is never certain, and foresight is never perfect, even for historical, transformative creators (Simonton, 2012). But those who are active participants in creating the transformation have a higher likelihood of influencing the trajectory. It is a shame that more creators do not see this forecasting as a responsibility. For example, interviews with veteran geneticists found that they tended not to consider how their findings might be used later (Gardner, Csikszentmihalyi, & Damon, 2001). Younger apprentices were similarly short-sighted, emphasizing their need to "get ahead" now, whereas ethics could wait until later (Fischman et al., 2004). Only a few "trustees" looked beyond their own professional needs to consider, and perhaps guide, the field's influence within the wider society and toward a greater good (Gardner, 2009).

One way to feel less anxious and more productive may be to improvise our way through the uncertainty. Improv—slang for improvisation—is a method of performance whose one and only rule is "yes, and…" Take what the situation—and other players within the situation—give us, then add our own spin. Thus, improv combines the contribution, consideration, and kaizen proposed earlier. We must move the scene forward, so we are contributing; we cannot reject or denounce what others have brought to the stage, so consideration is a must; and we must propel the plot forward, which is kaizen for the future.

Some individuals cultivate these skills so well that they become dispositions. Creativity—including in ethics—becomes proactive. They do not only foresee creative opportunities, but anticipate them. They position themselves not *for* the future, but *into* the future. They are "one step ahead." These dispositions particularly prevail in "problem finders," individuals who initiate new approaches with little stimulus from the environment (Getzels & Csikszentmihalyi, 1976; see also Kaufmann, 2004; Hersh, 1990, provides an example of proactive, anticipatory problem-finding regarding ethics in a creative field, mathematics).

By critically examining how creativity and ethics not only influence each other over time, but compose each other over time, we can gain insights regarding when in the creative process ethical issues are more likely to be addressed (Lloyd, 2009; Mumford et al., 2010) and what opportunities may support a positive role for creativity in ethical decision-making (Thaler & Sunstein, 2008). Kampylis and Valtanen (2010) warn that it is insufficient to recognize creativity's "malevolent/destructive and benevolent/constructive aspects" (p. 208). We must integrate creativity and ethics to address the consequences we create.

Two examples: Wikipedia and literature

What do these issues at the intersection of creativity and ethics look like in vivo, as individuals and communities are making judgments about new ideas and products as they become aware of them? Two examples from recent studies show the contrast between the perspective of the adopter (or non-adopter), the people who receive the novelty, and the perspective of the creator or innovator, the people who initiate the novelty.

Many people consider Wikipedia to be a creative endeavor, as millions of people worldwide add their contributions to the online, ever-changing encyclopedia. Yet, Wikipedia's culture stresses relatively strict adherence to a priori rules based on consensus (http://en.wikipedia. org/wiki/Wikipedia:Policies_and_guidelines). Its "pillars" specify that contributions (1) have a neutral point of view and not espouse a particular perspective, which usually means already socially accepted perspectives are more welcome than less common perspectives; (2) be verifiable from a reliable external source; and (3) not include novel claims, speculation, or unsubstantiated syntheses—"no original research" is allowed. In other words, Wikipedia editors are limited to mimicking or reworking what is already available in other reputable publications or what has already been accepted within Wikipedia (Moran, 2011).

As a default, novelty is considered corrupting. The community is more concerned with a "bad" edit staying on a page than a "good" edit introduced on a page. Thus, as a policy, Wikipedians are resistant to potentially creative edits for which, at the point the ideas are contributed, they are unsure whether the edit is "right" or "wrong." Error correction is more valued than interesting or provocative ideas.

The Wikipedia article-building process is like a scavenger hunt for new bits of information. Once in a while, an editor may call for organization, reorganization, or culling. Only occasionally does someone come along to integrate material, consider implications, or suggest an alternative view. A study of more than 10,000 edits across several controversial Wikipedia articles found that 98 percent of vandalism, 87 percent of misinformation, and 75 percent of spam was removed within a day, many on the next edit. Judgments of such "bad" contributions were quick. Judgments of contributions that inject information that may be "good," on the other hand, may take months as editors provide justifications on the discussion page and come to consensus (Moran, 2011). In this sense, Wikipedia is not much different from other institutional settings: it is more willing to avoid loss than to make a gain—what could go *wrong* is scarier than what could go *right* (Kahneman, Slovic, & Tversky, 1989).

However, Wikipedia makes visible possible reasons or implications for many people being ambivalent about creativity. Creativity requires flexibility not just from the originator but also from the receiving community. Rules devised out of fear produce more rote behavior and less situational sensitivity (Bierly, Kolodinsky, & Charette, 2009; Chonko, Wotruba, & Loe, 2003). Furthermore, when people can sense change as it is occurring, reasons and purposes are important for communication within a community. Removal of vandalism or spam from a Wikipedia page requires neither an edit summary nor a source citation. But contributing a new sentence may be eyed suspiciously if an explanatory edit summary or source citation is not forthcoming (Moran, 2011).

Wikipedia tends to take the perspective of the adopters, those who must receive and judge something unfamiliar. But a surprising finding that emerged from my study of commitment among highly success-ful novelists and poets (Moran, 2009b) emphasizes the perspective of the initiating innovator. In the study, writers were grouped, through a survey of the current literary field, into genre conformers whose work was competent and successful but similar to existing genre expecta-tions; experimentalists whose work was original and eccentric nearly to the point of being hard to comprehend; and domain transform-ers whose work revolutionized the literature canon. In terms of ethics, the genre conformers abided by the rules and norms set by authorities within the literary field and eventually took on those authority roles themselves. They were more conventional writers who tended to have forward-looking, stable, articulated moral purposes to be accurate and true to current standards and communities of writers. The experimen-talists, on the other hand, were the "rebels" who eschewed the field's or society's expectations. These writers tended to *draw on* conventional moral *supports*, but in a twisted, heretical way of standing "against" or "hiding from" conventions. Their work—such as beatnik stories or Henry Miller's sexually explicit novels—tended to be controversial, and their works were only moderately accepted by the literary field, earning them reputations as eccentric, self-centered, and often antisocial and immoral.

The domain transformers, such as Ernest Hemingway, William Faulkner, or Toni Morrison, who committed to some aspect of language or writing that they felt impassioned to share with readers, talked about being a vehicle for bringing the as yet untapped potential within the literature domain to readers' open minds. They spoke the most about a wider sense of responsibility. It was not the genre-conforming expert writers who abided by the ethical rules of the literary field, nor the experimental fringe writers who were labeled as rule-breakers, but rather the writers who eventually transformed the canon who spoke with the strongest sense of moral *purpose*. Through their works, they aimed to surprise the reader and change the reader's mind to a *wider* moral-ity beyond the current norm or standard. This drive to change minds carried a strong consideration of the effects of their works. But the pri-mary responsibility was not to other people directly. Rather, it was to change minds to support the writers' responsibility to the *domain*, to what language could contribute to the world.

Culture marches forward through both creativity and ethics

Everyone participates in culture, including its creation and re-creation (Glaveanu, 2011), whether we want to own up to that responsibility or not. On the one hand, those who contribute to the status quo may feel more supported, justified, and comfortable that what they are doing is "right" because the status quo is familiar and reinforced by cultural institutions. But institutions can support detrimental as well as worthy ends, such as slavery, warfare, discrimination, and so on. On the other hand, those who aim toward change may feel unsure or even ostracized, and must spend considerably more energy to not only pursue their chosen path, but blaze that path. "Rightness" must be created alongside the change; effective a priori formulations of the "good" in changing times are difficult to find (Bourdieu, 1993). Furthermore, creativity can entangle good and bad: creative successes, not failures, can produce the greatest harms alongside the greatest benefits (Tenner, 1996). Consider energy production and climate change, or cars and a host of social ills from urban sprawl to obesity to DWI (driving while impaired) crashes. Or, looking to the future, consider medical breakthroughs that increase lifespan to age 125 and may cause overpopulation.

We are back to ethical judgment—something to direct our actions. But we need ethical judgment beyond the given, with foresight and anticipation, that aims to enlarge our possibilities to contribute to the greater good. We need ethical judgment that expands our ethical horizons. Creators who want to strongly influence the direction of their fields, and society in general, *should* anticipate consequences they can bring about. As Winston Churchill said, "The price of greatness is responsibility."

Ethics shows that what creators *do* matters. Creativity creates a bumpier ride: the result is more unpredictable than if the situation is the status quo, when we can count on tomorrow to be much like today was. Our optimism aims to improve our world, but the law of unintended consequences warns us our intentions could backfire. We need as much creativity in addressing ethical quandaries as we put into designing innovations, and we need creative ethical thinking to come full circle to address the ethical implications of the innovations we launch into the world. Then we will have an ethics of possibility.

References

Aristotle. (1999). *Nicomachean Ethics*, 2nd ed. (T. Irwin, trans.). Cambridge, MA: Hackett Publishing.

Baughman, W. A., & Mumford, M. D. (1995). Process-analytic models of creative capacities: Operations influencing the combination-and-reorganization process. *Creativity Research Journal, 8*(1), 37–62.

Becker, H. S. (1963). *Outsiders: Studies in the Sociology of Deviance.* New York: Free Press.

Bierly, P E. III, Kolodinsky, R. W., & Charette, B. J. (2009). Understanding the complex relationship between creativity and ethical ideologies. *Journal of Business Ethics, 86*, 101–112.

Blasi, A. (1980). Bridging moral cognition and moral action: A critical review of the literature. *Psychological Bulletin, 88*, 1–45.

Bourdieu, P. (1993). *The Field of Cultural Production.* New York: Columbia University Press.

Bruner, J. S. (1962). The conditions of creativity. In J. S. Bruner, *On Knowing: Essays for the Left Hand.* New York: Cambridge University Press.

Byrne, C. L., Shipman, A. S., & Mumford, M. D. (2010). The effects of forecasting on creative problem-solving: An experimental study. *Creativity Research Journal, 22*(2), 119–138.

Cho, M. (2010). Envisioning Seoul as a world city: The cultural politics of the Hong-dae Cultural District. *Asian Studies Review, 34*, 329–347.

Chonko, L. B., Wotruba, T. R., & Loe, T. W. (2003). Ethics code familiarity and usefulness: Views on idealist and relativist managers under varying conditions of turbulence. *Journal of Business Ethics, 42*(3), 237–252.

Christensen, C., Craig, T., & Hart, S. (2001). The great disruption. *Foreign Affairs*, 80–95.

Cremin, T., Burnard, P., & Craft, A. (2006). Pedagogy and possibility thinking in the early years. *Thinking Skills and Creativity, 1*(2), 108–119.

Csikszentmihalyi, M. (1988). Society, culture, and person: A systems view of creativity. In R. J. Sternberg (ed.), *The Nature of Creativity* (pp. 325–339). New York: Cambridge University Press.

Csikszentmihalyi, M., & Nakamura, J. (2007). Creativity and responsibility. In H. Gardner (ed.), *Responsibility at Work* (pp. 64–80). San Francisco, CA: Jossey-Bass.

Cua, A. S. (1978). *Dimensions of Moral Creativity: Paradigms, Principles, and Ideals.* University Park: Pennsylvania State University Press.

Darnton, R. (2009). Google and the future of books. *The New York Times Review of Books*, February 12, *56*(2). http://www.nybooks.com/articles/archives/2009/feb/12/google-the-future-of-books/, accessed August 7, 2013.

David-Ferdon, C., & Hertz, M. F. (2007). Electronic media, violence, and adolescents: An emerging public health problem. *Journal of Adolescent Health, 41*(6), S1–S5.

Dudley, W. C. (2010). Asset bubble and the implications for central bank policy. Speech at Economic Club of New York, New York, April 7. http://www.bis.org/review/r100409c.pdf, accessed November 11, 2013.

Edwards, C. P. (1981). The comparative study of the development of moral judgment and reasoning. In R. H. Munroe, R. L. Munroe, & B. B. Whiting (eds.), *Handbook of Cross-Cultural Human Development* (pp. 501–528). New York: Garland Press.

European Commission. (2009). *The Impact of Culture on Creativity.* Report of the Directorate-General for Education and Culture. Brussels: European Commission.

Feldman, D. H., Csikszentmihalyi, M., & Gardner, H. (1994). *Changing the World: A Framework for the Study of Creativity*. Westport, CT: Praeger.

Fischman, W., Solomon, B., Greenspan, D., & Gardner, H. (2004). *Making Good: How Young People Cope with Moral Dilemmas at Work*. Cambridge, MA: Harvard University Press.

Florida, R. L. (2002). *The Rise of the Creative Class and How It's Transforming Work, Leisure, Community, and Everyday Life*. New York: Basic Books.

Florida, R. L. (2012). *The Rise of the Creative Class: Revisited*. New York: Basic Books.

Gardner, H. (2009). I. What is good work? II. Achieving good work in turbulent times (G. B. Petersen, ed.). *The Tanner Lectures on Human Values, 28*, 199–233.

Gardner, H., Csikszentmihalyi, M., & Damon, W. (2001). *Good Work: When Excellence and Ethics Meet*. New York: Basic Books.

Getzels, J., & Csikszentmihalyi, M. (1976). *The Creative Vision*. New York: John Wiley & Sons.

Gibbs, J. C. (2003). *Moral Development and Reality: Beyond the Theories of Kohlberg and Hoffman*. New York: Sage.

Gigerenzer, G. (2010). Moral satisficing: Rethinking moral behavior as bounded rationality. *Topics in Cognitive Science, 2*(3), 528–554.

Gilligan, C. (1982). *In a Different Voice*. Cambridge, MA: Harvard University Press.

Glaveanu, V. P. (2011). Creativity as cultural participation. *Journal for the Theory of Social Behaviour, 41*(1), 48–67.

Granovetter, M. (1983). The strength of weak ties: A network theory revisited. *Sociological Theory, 1*, 201–233.

Gruber, H. E. (1993). Creativity in the moral domain: Ought implies can implies create. *Creativity Research Journal, 6*(1–2), 3–15.

Guilford, J. P. (1950). Creativity. *American Psychologist, 5*, 444–454.

Haidt, J. (2001). The emotional dog and its rational tail: A social intuitionist approach to moral judgment. *Psychological Review, 108*, 814–834.

Haidt, J., & Graham, J. (2007). When morality opposes justice: Conservatives have moral intuitions that liberals may not recognize. *Social Justice Research, 20*, 98–116.

Haidt, J., Koller, S. H., & Dias, M. G. (1993). Affect, culture, and morality, or is it wrong to eat your dog? *Journal of Personality and Social Psychology, 65*(4), 613–328.

Hall, J., & Rosson, P. (2006). The impact of technological turbulence on entrepreneurial behavior, social norms and ethics: Three internet-based cases. *Journal of Business Ethics, 64*, 231–248.

Haste, H., & Abrahams, S. (2008). Morality, culture and the dialogic self: Taking cultural pluralism seriously. *Journal of Moral Education, 37*(3), 377–394.

Hersh, R. (1990). Mathematics and ethics. *Mathematical Intelligencer, 12*, 13–15.

Jaques, E. (1955). Social systems as a defense against persecutory and depressive anxiety. In M. Klein, P. Heimann, & R.E. Money-Kyrle (eds.), *New Directions in Psychoanalysis* (pp. 478–498). London: Tavistock.

Kagan, J. (2009). Categories of novelty and states of uncertainty. *Review of General Psychology, 13*(4), 290–301.

Kahnemann, D., Slovic, P., & Tversky, A. (1989). *Judgment under Uncertainty: Heuristics and Biases*. New York: Cambridge University Press.

Kampylis, P. G., & Valtanen, J. (2010). Redefining creativity—Analyzing definitions, collocations, and consequences. *Journal of Creative Behavior, 44*(3), 191–214.

Kaufmann, G. (2004). Two kinds of creativity—but which ones? *Creativity and Innovation Management, 13*(3), 154–165.

Kirton, M. (1976). Adaptors and innovators: A description and measure. *Journal of Applied Psychology, 61*, 622–629. doi:10.1037/0021-9010.61.5.622.

Kohlberg, L. (1981). *Essays on Moral Development, Vol. I: The Philosophy of Moral Development*. New York: Harper & Row.

Krebs, D. L., & Denton, K. (2005). Toward a more pragmatic approach to morality: A critical evaluation of Kohlberg's model. *Psychological Review, 112*(3), 629–649.

Liu, Y., Gummadi, K. P., Krishnamurthy, B., & Mislove, A. (2011). Analyzing Facebook privacy settings: User expectations vs. reality. *Proceedings of the 2011 ACM SIGCOMM Conference on Internet Measurement*, 61–70. doi:10.1145/2068816.2068823.

Lloyd, P. (2009). Ethical imagination and design. *Design Studies, 30*, 154–168.

Lopes, P. (2005). Signifying deviance and transgression: Jazz in the popular imagination. *American Behavioral Scientist, 48*(11), 1468–1481.

Magedley, W., & Birdi, K. (2009). Innovation labs: An examination of the use of physical spaces to enhance organizational creativity. *Creativity and Innovation Management, 18*(4), 315–325.

Marks, E. (2002). *Business Darwinism, Evolve or Dissolve: Adaptive Strategies for the Information Age*. New York: John Wiley & Sons.

Maslow, A. H. (1973). Creativity in self-actualizing people. In A. Rothenberg & C. R. Hausman (eds.), *The Creative Question* (pp. 86–92). Durham, NC: Duke University Press.

McLaren, R. (1993). The dark side of creativity. *Creativity Research Journal, 6*, 137–144.

Mina, M., Kemmet, S., & Gerdes, R. (2007). Work in progress—the next step for cubism in education: From abstract concepts to a classroom based implementation. Frontiers in Education Conference, Milwaukee, WI, October 10–13.

Moran, S. (2009a). Creativity: A systems perspective. In T. Richards, M. Runco, & S. Moger (eds.), *The Routledge Companion to Creativity* (pp. 292–301). London: Routledge.

Moran, S. (2009b). What role does commitment play among writers with different levels of creativity? *Creativity Research Journal, 21*(2–3), 243–257.

Moran, S. (2010a). Changing the world: Tolerance and creativity aspirations among American youth. *High Ability Studies, 21*(2), 117–132.

Moran, S. (2010b). Creativity in school. In K. S. Littleton, C. Wood, & J. K. Staarman (eds.), *International Handbook of Psychology in Education* (pp. 319–360). Bingley: Emerald Group.

Moran, S. (2010c). Returning to the Good Work Project's roots: Can creative work be humane? In H. Gardner (ed.), *Good Work: Theory and Practice* (pp. 127–152). Cambridge, MA: Good Work Project.

Moran, S. (2010d). The roles of creativity in society. In J. C. Kaufman & R. J. Sternberg (eds.), *The Cambridge Handbook of Creativity* (pp. 74–90). New York: Cambridge University Press.

Moran, S. (2011). Creativity for the greater good: How people decide new ideas are morally good or corrupt. Washington, DC: American Psychological Association Convention.

Mumford, M. D., & Gustafson, S. B. (1988). Creativity syndrome: Integration, application, and innovation. *Psychological Bulletin, 103*(1), 27–43.

Mumford, M. D., Waples, E. P., Antes, A. L., Brown, R. P., Connelly, S., Murphy, S. T., & Devenport, L. D. (2010). Creativity and ethics: The relationship of creative and ethical problem-solving. *Creativity Research Journal, 22*(1), 74–89.

Oakley, K. (2004). Not so cool Britannia: The role of the creative industries in economic development. *International Journal of Cultural Studies, 7*, 67–77.

Patrick, K., Griswold, W. G., Raab, F., & Intille, S. S. (2008). Health and the mobile phone. *American Journal of Preventative Medicine, 35*(2), 177–181.

Polanyi, M. (1958). *Personal Knowledge: Towards a Post-Critical Philosophy.* Chicago, IL: University of Chicago Press.

Pratt, A. C., & Jeffcutt, P. (eds.) (2009). *Creativity, Innovation and the Cultural Economy.* New York: Routledge.

Rawls, J. (1971). *A Theory of Justice.* Cambridge, MA: Belknap Press.

Rawls, J. (2001). *Justice as Fairness: A Restatement.* Cambridge, MA: Belknap Press.

Rest, J. R. (1986). *Moral Development: Advances in Research and Theory.* New York: Praeger.

Rogers, E. M. (1983[1962]). *Diffusion of Innovations*, 3rd edn. New York: Free Press.

Rozin, P., Lowery, L., Imada, S., & Haidt, J. (1999). The CAD triad hypothesis: A mapping between three moral emotions (contempt, anger, disgust) and three moral codes (community, autonomy, divinity). *Journal of Personality and Social Psychology, 76*, 574–586.

Runco, M. A., & Nemiro, J. (2003). Creativity in the moral domain: Integration and implications. *Creativity Research Journal, 15*(1), 91–105.

Runco, M., & Richards, R. (eds.) (1989). *Eminent Creativity, Everyday Creativity, and Health.* Greenwich, CT: Ablex.

Ryder, K. (2011). Singapore's government gets what it wants—from social policy to tourism. But its bid to enhance the nation's creative atmosphere may be a challenge more difficult any other. *Fortune*, September 13. http://tech.fortune. cnn.com/2011/09/13/singapore-startups/, accessed November 11, 2013.

Schumpeter, J. A. (1939). *Business Cycles: A Theoretical, Historical and Statistical Analysis of the Capitalist Process.* London: McGraw-Hill.

Scott, G. M., Lonergan, D. C., & Mumford, M. D. (2005). Conceptual combination: Alternative knowledge structures, alternative heuristics. *Creativity Research Journal, 17*(1), 79–98.

Shweder, R. A., Mahapatra, M., & Miller, J. G. (1987). Culture and moral development. In J. Kagan & S. Lamb (eds.), *The Emergence of Morality in Young Children* (pp. 1–83). Chicago, IL: University of Chicago Press.

Simonton, D. K. (1994). *Greatness: Who Makes History and Why.* New York: Guilford Press.

Simonton, D. K. (2004). *Creativity in Science: Chance, Logic, Genius, and Zeitgeist.* New York: Cambridge University Press.

Simonton, D. K. (2012). Foresight, insight, oversight, and hindsight in scientific discovery: How sighted were Galileo's telescopic sightings? *Psychology of Aesthetics, Creativity, and the Arts, 6*(3), 243–254.

Stebbins, R. A. (1966). Class, status, and power among jazz and commercial musicians. *Sociological Quarterly, 7*(1), 197–213.

Stenmark, C. K., Antes, A. L., Wang, X., Caughron, J. J., Thiel, C. E., & Mumford, M. D. (2010). Strategies in forecasting outcomes in ethical decision-making: Identifying and analyzing the causes of the problem. *Ethics and Behavior, 20*(2), 110–127.

Tappan, M. B. (1997). Language, culture, and moral development: A Vygotskian perspective. *Developmental Review, 17*(1), 78–100.

Tenner, E. (1996). *Why Things Bite Back: Technology and the Revenge of Unintended Consequences.* New York: Vintage.

Thaler, R. H., & Sunstein, C. R. (2008). *Nudge: Improving Decisions about Health, Wealth, and Happiness.* New Haven, CT: Yale University Press.

Thomson, J. J. (1976). Killing, letting die, and the trolley problem. *The Monist, 59*, 204–217.

Tuomi, I. (2001). Internet, innovation, and open source: Actors in the network. *First Monday*, 6(1). http://firstmonday.org/ojs/index.php/fm/article/view/824/733, accessed August 7, 2013.

Volkow, N. D., Tomasi, D., Wang, G. J., Vaska, P., Fowler, P., et al. (2011). Effects of cell phone radio frequency signal exposure on brain glucose metabolism. *JAMA: Journal of the American Medical Association, 305*(8), 808–813.

Vygotsky, L. S. (1978). *Mind in Society: The Development of Higher Psychological Processes* (M. Cole, V. John-Steiner, S. Scribner, & E. Souberman, eds.). Cambridge, MA: Harvard University Press.

Welch, E., Bretschneider, S., & Rohrbaugh, J. (1998). Accuracy of judgmental extrapolation of time series data: Characteristics, causes, and remediation strategies for forecasting. *International Journal of Forecasting, 14*, 95–110.

Wertheimer, M. (1945). *Productive Thinking.* New York: Harper & Row.

Weston, A. (2006). *Creative Problem-Solving in Ethics.* New York: Oxford University Press.

Whitworth, M. (2002). *Einstein's Wake: Relativity, Metaphor, and Modernist Literature.* New York: Oxford University Press.

Summary

Creativity and Ethics—Two Golden Eggs

David Cropley
University of South Australia, Australia

James C. Kaufman
University of Connecticut, USA

Michelle Murphy
University of South Australia, Australia

and

Seana Moran
Clark University, USA

> *Once, as a little girl, she had dreamed that she was locked inside a golden egg that flows through the universe. Everything was pitch-black, there weren't even any stars, she'd have to stay there forever, and she couldn't even die. There was only one hope. Another golden egg was flying through space. If it collided with her own, both would be destroyed, and everything would be over. But the universe was so vast!*
>
> Tim Krabbé, 1993, pp. 13–14

The quote above is taken from the novel *The Vanishing* (later adapted into a successful Dutch film and less successful American film). The main character, Saskia, describes her recurring dream of loneliness. Academia can tend toward such isolation, with different disciplines and topics staying within their own "golden eggs" of discourse. One of the goals of a volume such as this one is to force two "golden eggs" to collide by asking scholars from the fields of both creativity and ethics to consider how these topics might be related.

Blending creativity with its consequences makes the generation of novel and effective solutions more than merely a question of what we could do, and turns it as much into an issue of what we should do. Crang and Cook (2007, p. 5), writing about the discipline of ethnography,

299

emphasized that people's actions are "the result of individuals drawing on the structure of their culture" and not simply responding to what the environment offers. The ethics of creativity shines a similar light on the development of new ideas, processes, services, and products. What effect does creativity have on individuals, groups, and societies? How do the fundamental values on which they base their actions and institutions play a role in creativity? What constitutes good and evil, right and wrong, and how does creativity "rock the boat" about these beliefs?

The growth of interest in the dark side of creativity over the last few years has stimulated a deeper examination of these questions. Is creativity universally and unequivocally good, or can it be misapplied in the service of negative, or indeed deliberately harmful, ends? Although criminally deviant behavior is perhaps the most visible tip of the iceberg, it is in the more ambiguous realm of morality and ethical behavior, found across disciplines as diverse as teaching and engineering, that our exploration needs to venture. Why, for example, do most people espouse the virtues of creativity yet frequently quash it in themselves and others? How can creative solutions to problems also give rise to new ethical problems? When and how do creativity and ethics come into conflict, and how can the two co-exist and reinforce each other in practical and useful ways?

The book containing the opening quote provides an example of the often twisty relationship between creativity and ethics. Saskia vanishes, and her boyfriend Rex is so driven to find out what happens to her that he is willing to give up anything simply to find out. He meets Lemorne, the man responsible for her disappearance. Despite Lemorne's questionable morality, he believes it is unethical not to offer Rex a chance to find out the truth. Lemorne derives a creative solution: Rex can find out what happened to Saskia if Rex is willing to submit himself to the identical fate. The ultimate decision of whether to find out (and, perhaps, of life and death) is left up to Rex. Lemorne is highly creative and acts in a way that an outside observer would consider immoral; yet, he maintains his own personal standard of ethics.

To tackle similarly vexing situations at the crossroads of ethics and creativity, we focused on three questions to structure the discussion. Here, we will briefly summarize our contributors' key points.

The development of ethical creativity

Part I addressed the question: What are the moral mental mechanisms involved in creativity, and how do they develop? Five authors put forth arguments that these mechanisms involve imagination, embodied know-how, reasoning, discernment, and integrity. In "The Development of Moral Imagination" Darcia Narvaez and Kellen Mrkva discuss the development of *creative moral imagination*. Defining *moral imagination* as "not only the ability to generate useful ideas, but also abilities to form ideas about what is good and right, and to put the best ideas into action for the service of others," they outline how a person's imagination, emotion, and morality develop through experience to form four types of moral imagination: *detached* (little emotional engagement with the world), *vicious* (aggressive emotional interaction with the world), *engaged* (present-focused positive interaction with the world), and *communal* (extended, collaborative, positive interaction with the world). The authors suggest that there are different amounts of creativity, and a different *quality* of creativity, within each of these four types of moral imagination.

In "Moral Craftsmanship" Mark Coeckelbergh suggests that individuals learn from others how to be moral via skilled instruction and social engagement. However, eventually, these individuals learn to "craft" their own unique twist on their moral compass using both imagination and creativity, rather than following the moral rules provided by society. Hence, Coeckelbergh proposes that individuals develop their morality through the concept of *moral craftsmanship*.

In "Creativity in Ethical Reasoning" Robert J. Sternberg argues that it is hard to be ethical without being creative. Whereas high intelligence does not necessarily lead to more ethical behavior, creativity is a decision that is necessary for ethical reasoning. Nonetheless, creativity and ethics do not always go together; for example, Stanley Milgram's studies of obedience and Phil Zimbardo's studies of prisoners and guards, as well as unethical government actions, such as the Nazi genocide, the Armenian genocide, and the Rwandan genocide, show that individuals and groups can be quite creative in their behavior affecting others without recognizing the harm they inflict. Sternberg argues that schools should teach how to think about issues of right and wrong through *case scenarios* by which students can practice creative solutions to ethical dilemmas.

In "Moral Creativity and Creative Morality" Qin Li and Mihaly Csikszentmihalyi describe "ethical anarchists"—individuals over 60 years old, including some Nobel Prize winners, who spent their lives in professions requiring creativity. Rather than considering morality as something universal to everyone, ethical anarchists ignore conventional morality and strongly endorse professional ethics. These ethics tied to their work give them a moral grounding in integrity.

The Ethics of Creativity

In "Creative Artists and Creative Scientists: Where Does the Buck Stop?" James Noonan and Howard Gardner suggest there are consequences to creative activity, which they call "post-creative developments." All creative individuals—whether they actively seek out post-creative ethical entailments, or the ethical issues find them—must decide how they would like to respond. Noonan and Gardner describe individuals who represent four types of post-creative development roles: the *Opportunist*, portrayed through artist Shepard Fairey's famous Obama poster; the *Reluctant Winner*, exemplified by songwriter Gretchen Peters' passive benefits from others' misuse of her hit "Independence Day"; the *Unlucky Gambler*, characterized by how Nazi-era German filmmaker Leni Riefenstahl's film *The Triumph of Will* became a model of Nazi propaganda, which she then used to distort the influence of Hitler and Nazism on her work, which undermined her reputation; and the *Hostage*, represented by Erich Maria Remarque's war novel *All Quiet on the Western Front*, which went from instant international sensation to banned book. Whereas professionals tend to be guided by professional ethics, creative non-professionals rely on informal social networks and individual discernment for their ethics. Hence, there is likely to be greater variability in how non-professionals respond when confronted with ethical dilemmas. Noonan and Gardner conclude that creative individuals are not responsible solely *for* ideas, but are also responsible *to* ideas. This responsibility extends well beyond the moment at which individuals put their ideas into the world.

The conditions for ethical and unethical creativity

Part II addressed the question: When, how, and why does creativity lead to positive or negative ethical impacts—or both? Five authors consider how aspects of creativity—such as personality, identity, freedom and constraints of the problem, problem construction, and technology—can affect the ethics of the work process or outcome.

In "A Creative Alchemy" Ruth Richards presents a positive, optimistic view that being creative can make us more authentic human beings, becoming more moral through creativity's beneficial effects, which ultimately helps society as well. Creativity can be used for moral good. Creative behaviors lead to improved well-being, and healing arts and expressive arts therapies can help alleviate suffering. She challenges us to consider how creativity can "take us to a more open, authentic, aware, non-defensive, and also attuned, connected, caring, and morally elevated place where we can live better, individually and together."

In "License to Steal: How the Creative Identity Entitles Dishonesty" Lynne C. Vincent and Jack A. Goncalo discuss recent research showing a relationship between creative identity—seeing oneself as creative— and dishonesty. Individuals who maintain a creative identity, as well as the people who interact with these individuals, tend to believe that creative individuals are entitled to engage in immoral behavior. One reason is that creativity is associated with negative traits, such as self-interestedness, challenging the status quo, and breaking rules. Another reason is that these individuals believe they justly deserve more than others, and are willing to engage in unethical behaviors to gain the rewards they believe they deserve. This unethical situation can be perpetuated at the organizational level. Organizations that support creativity tend also to support the sense of entitlement in creative individuals. In turn, when organizations act against these individuals' sense of entitlement, the individuals can lash out by engaging in dishonest behavior. To prevent dishonest behavior, organizations must devalue entitled behaviors while still upholding values that support their employees' creativity.

In "Engineering, Ethics, and Creativity: N'er the Twain Shall Meet?" David H. Cropley reiterates that creativity has been associated with courage, openness to experience, flexibility, and more healthful mental states, yet there is also the potentially *dark side* of creativity, which he addresses particularly in engineering design. Classifying negative *outcomes* of faulty engineering designs as either benevolent or malevolent creativity can be difficult. What is important is to consider the ethics associated with the *development* of the technology. There is an apparent paradox between the unconstrained nature of creative thought and the constraints of technological and engineering design rules. When designing new products to keep their organizations competitive, engineers also need to consider the safety of and risk to users and others. Yet, innovation and rules need not be antagonistic: constraints inherent to engineering design are a potential stimulus for creativity. Utilitarianism, duty, and virtue all can play a role in creative problem-solving. Freedom and constraint in the context of engineering design are reconcilable, and it is entirely possible to devise ethical *and* creative engineering solutions.

In "Construction or Demolition: Does Problem Construction Influence the Ethicality of Creativity?" Daniel J. Harris, Roni Reiter-Palmon, and Gina Scott Ligon suggest that the way creative people construct problems, based on representations of their past experiences, can greatly affect how they address ethical issues. By examining life experiences of historically notable and destructive individuals and the ethicality of decisions they made, the authors showed how leaders who had early experiences with negative life themes, object beliefs, and personalized need for power were more likely to make impetuous decisions with destructive consequences. These past experiences and life themes of negative events and destructiveness would lead to justification mechanisms and situational construal that promotes aggression. The cognitive and developmental antecedents of malevolent creativity require further study, especially considering the harm that may result from the actions of those who construct their realities in highly destructive ways.

In "Intelligent Decision-Making Technology and Computational Ethics" Anthony Finn examines how the increase in technology of microprocessors and software has led to an ability to create autonomous and intelligent decision-making technologies, such as robots or unmanned aerial vehicles (drones), that can change our lives radically. Although there are many limitations that must be overcome before these technologies will have human-like reasoning, they shift from being tools that humans use into being service providers that overcome human frailties. However, errors produced by these technologies show how humans are no longer in control because the machines themselves can synthesize and execute their own decision-making processes. Finn suggests that there are several ways developers can constrain machines to *behave* ethically, but for machines to make ethical *judgments* is more complicated. Still, considerable resources are invested in improving hardware and algorithms in the hope of technologies that better serve humanity. As intelligent decision-making technologies become more creative, the challenge to code software capable of resolving complex dilemmas with ethical dimensions will become more pronounced.

Ethics in creativity: Supporter or obstacle?

Part III addresses the question: What role does ethics play in supporting or thwarting creativity? Five chapters examine how aspects of ethical constraints or decision-making affect creative work in five domains: arts, sciences, photography, education, and leadership.

In "Creative Transformations of Ethical Challenges" Vera John-Steiner and Reuben Hersh explore the ethics of choices that accompany creative contributions through three historical case studies of individuals who demonstrate exemplary creative altruism in the midst of challenging circumstances. Physicist Leo Szilard, who patented the concept of a nuclear reactor and worked on the Manhattan Project in the Second World War, faced the ethical dilemma between the threat of Nazism and Japanese imperialism and the terrible destruction and loss of innocent life that could be unleashed by an atomic bomb. Rather than denying the horrific results of his discoveries, Szilard devoted much of his life to seeking ethical and creative ways to reshape those consequences. Painter Friedl Dicker-Brandeis, who taught art to children in the Terezín concentration camp, faced the ethical dilemma between being used in German propaganda as an example of positive activities provided to prisoners and helping children deal with their hardships. Friedl saved more than 5,000 pieces of the children's artwork, which were given to the Jewish Museum in Prague. African American mathematics professor Clarence Stephens faced the ethical dilemma between students not enjoying and failing mathematics through traditional methods and launching a radically different college educational approach that was untested. Stephens believed that any college student could learn mathematics given the right conditions.

In "The Hacker Ethic for Gifted Scientists" Kirsi Tirri suggests that gifted scientists are hackers, people who work based on intrinsic interest and enjoyment. Using case studies of educational tasks for gifted science students and professional researchers, she shows how a "hacker ethic" based on an enjoyment of doing good works can best educate creative, moral people in science. High intellectual ability does not predict mature moral judgment, which also requires moral motivation and moral sensitivity. Thus, from an early age, budding scientists must be sensitized to—and motivated to engage—the ethical aspects of scientific work. Through explicit discussion of ethical quandaries and social and moral issues stemming from science, ethics can be tied to scientists' inner drive to excel and create.

In "The Dialogic Witness: New Metaphors of Creative and Ethical Work in Documentary Photography" Charlotte Dixon and Helen Haste argue that documentary photography suffers from a conflict between innovative, affordable cameras that allow anyone to snap a shot of personal and civic importance, and the realism of photography that denies the role of subjective judgment in what is being recorded. Photographers such as Jacob Riis, Lewis Hine, Eugene Smith, and Sebastio Salgado have left enduring legacies in social reform as their photographs helped alter social attitudes. Thus, they engaged in moral creativity, which involves seeing what others have failed to identify, understanding the importance of acting, and recognizing the outcomes of acting on one's vision.

In "Neglect of Creativity in Education: A Moral Issue" Arthur Cropley addresses society's general aversion to creativity as an ethical issue. Although teachers value creativity, they tend to not like creative students. Implicitly, this aversion sends signals to students to be afraid of creativity and its associated uncertainty, which, over the long term, may undermine their capacities to face ethical issues that require creativity. Cropley argues that educational institutions and educators have a moral responsibility to promote creativity because it has the potential to make vital contributions to the common good. They must address creativity's "bundle of paradoxes" that stem from a conventional view of morality, including creativity as breaking rules, challenging authority, creating conflict, or taking risks.

In "The Ethical Demands Made on Leaders of Creative Efforts" Michael Mumford and colleagues note how creative people may not engage in creative work when they perceive the environment to be hostile. A key role for leaders is to provide an environment conducive to creative work, yet also highlight how the work done can affect others beyond the creative team. Creative efforts complicate leadership because of ill-defined problems, uncertainty, and the team's demands for autonomy. At its core, ethics-oriented leadership for creative efforts involves sensemaking, devising a coherent story by which the team, the organization, and affected others can understand the work process, including forecasting its consequences. This sensemaking goes beyond technical issues to incorporate the social issues at play, recognizing that the creative team is embedded within larger social systems.

Horizons

In "An Ethics of Possibility" Seana Moran looks to the future: How can creativity drive a view of ethics that is not limited to rules or tradition, but rather embraces opportunities with anticipation to make societies and cultures richer, stronger, and better able to perceive larger effects in real time, not only retrospectively? Taking a cultural-historical view that posits both creativity and morality as endogenous forces that change cultures over time, she veers away from an either–or mentality: neither *all* difference/change/newness through incessant creativity nor *all* tradition/stability through a rigid ethics is good for cultures or individuals. Rather, since *each* of us contributes to our cultures in both stabilizing and creative ways, *all* of us should strive to "live on the edge," straddling what our culture has already accepted and what is possible and yet to manifest. Creative ideas, products, and so on are not "one-offs" but instead can start a ripple effect. What we do matters.

Revisiting the golden eggs

If we consider the fields of creativity and ethics to be their own golden eggs, then there are several perspectives presented in this book. Some essays discuss the merging of these two constructs into one, such as Narvaez and Mrkva's moral imagination, John-Steiner and Hersh's creative altruism, and Dixon and Haste's moral creativity. Other essays argue that one construct supports the other. Sternberg and Coeckelbergh discuss how the practice of being ethical entails creative effort, and Richards highlights how the act of being creative can improve one's ethics. Others hold that the constructs may impair each other. Vincent and Goncalo write how creative identity may be linked to a more flexible morality, and Li and Csikszentmihalyi consider the creative genius as ethical anarchist. David Cropley views ethics as a possible constraint on creativity, much like technical requirements.

Another approach is to discuss key determinants of both constructs, such as Harris, Reiter-Palmon, and Ligon's analysis of problem construction and Finn's breakdown of technical capacities. Some contributors use a specific field as a lens to examine the two constructs, such as education, documentary photography, engineering, and leadership. Finally, both Moran and Noonan and Gardner discuss how creative actions themselves can lead to moral and ethical repercussions.

What creativity and ethics may share the most is that both are difficult. Both take effort, care, and often additional work. Both can mean bucking the path of least resistance and standing up to others who may view such decisions as silly or a waste of energy. But, as history has shown through cases of moral and ethical exemplars with creative approaches to perplexing issues, facing such difficulty— when done thoughtfully, with respect for others, and recognizing the interconnections among us all—can be crucial for the greater good.

References

Crang, M., & Cook, I. (2007). *Doing Ethnographies*. Thousand Oaks, CA: Sage.
Krabbé, T. (1993). *The Vanishing*. (C. N. White, trans.). New York: Random House.

Index

Index Note: Page numbers in *italics* refer to figures and tables.

322 *Index*

Printed and bound in the United States of America